CHARLES REINER

LESSONS ON NUMBER
AS GIVEN
IN A PESTALOZZIAN SCHOOL

Elibron Classics
www.elibron.com

Elibron Classics series.

ISBN 1-4021-9764-0 (paperback)
ISBN 1-4021-2209-8 (hardcover)

This Elibron Classics Replica Edition is an unabridged facsimile of the edition published in 1851 by Taylor, Walton, and Maberly, London.

LESSONS ON NUMBER.

LESSONS ON NUMBER,

AS GIVEN

IN A PESTALOZZIAN SCHOOL,

CHEAM, SURREY.

THIRD EDITION, CORRECTED.

The Master's Manual.

By C. REINER,

TEACHER OF MATHEMATICS IN CHEAM SCHOOL.

LONDON:
TAYLOR, WALTON, AND MABERLY,
UPPER GOWER STREET, AND IVY LANE, PATERNOSTER ROW.

1851.

Preface to the Second Edition.

NUMBER presents a most important field, on which to develop and strengthen the minds of children. Its obvious connection with the circumstances surrounding them—the simplicity of its data—the clearness and certainty of its processes—the neatness and indisputable correctness of its results—adapt it in an eminent degree for early instruction. Arithmetical exercises tend to give clearness, activity, and tenacity to the mind; many an intellect that has not power enough for geometry, nor refinement enoegh for language, finds in them a department of study on which it may labour with the invigorating consciousness of success. But the advantages must of course depend, in a great measure, on the manner in which arithmetic is taught. More than any other branch of instruction has it suffered, in this country, from the influence of circumstances. The reproach that we are a nation of shopkeepers, might seem to have originated in the spirit of our arithmetical studies. Most popular treatises on the subject degrade the science they profess to elucidate; it is made a mere shop-boy's assistant, the *vade-mecum* of the counter and the desk. A certain mechanical dexterity in performing the operations of arithmetic, as required by the circumstances of commercial life, is effectually obtained by the use of these treatises; but the principles of the science are unknown, and many of its advantages, as presenting an exercise of mental power, altogether neglected.

Intelligent teachers, however, have not always been satisfied that their pupils should regard their mode of calculating as correct, or convenient, merely because it corresponded with the rule given in their book; they have explained the *rationale* of the process. The simple, lucid, and well-arranged treatise of Professor De Morgan, is among the happiest attempts to rescue arithmetic from its present degraded state, and to claim for it a place among other branches of rational education. It is peculiarly valuable for young persons, who, having been from their infancy led hood-winked through the dark alleys of arithmetical rules, desire to take an intellectual view of operations, which they have been taught to perform mechanically. It takes them, as it were, to an eminence, where they can see the point from which they started, and that at which they have arrived, and, tracing all the windings of the dark passages which they were made to traverse, shows them that they were indeed the shortest, if not the best course they could have followed.

The aim of the little work now offered to the public is different; it does not propose to explain processes, but to unfold principles. The pupil is not taught to comprehend a rule, but to dispense with it, or form it for himself. The path along which he is led may be longer than the usual route, but then it is in broad day-light; he is more independent of his guide, and derives more health and vigour from the exercise. Were the true ends of intellectual education more clearly apprehended, the means of prosecuting it would be more justly appreciated. While the question *cui bono?* so judicious in itself, is answered by a sordid reference to mere money-getting, or by a narrow-minded consideration of professional advancement, every method of instruction that proposes to itself a more exalted, though less obvious utility, will be ridiculed as visionary, or neglected as unprofitable. But when the true end of intellectual education shall be admitted to be, first, the attainment of mental power, and then the application of it to practical and scientific purposes, that plan of early instruction which dwells long on first principles, and does not haste to make learned, will be acknowledged as the most economical, because the most effectual. Experience will show, as indeed it has already shown, that while superficial teaching may prepare for the mere routine of daily business, whensoever a question, not anticipated in the manual, occurs, none but the pupil whose faculties have been exercised in the investigation of truth, who is the master, not the slave of rules, will solve the unexpected difficulty, by a novel application of the principles of the science.

Writers on method have observed that there is a certain order in which truths present themselves to the mind engaged in the original investigation of a subject, and that when the subject has been investigated, a different arrangement is necessary for the lucid exposition of the truths discovered. These views have been most unhappily applied in the early stages of instruction. For although the artificial order may be best calculated to convey knowledge to minds already trained for its reception, by previous acquaintance with similar subjects, it is by no means suited to the opening faculties of children. Hence the disgust, in many cases insurmountable, which the first principles of a science inspire in their minds. This disgust, however, vanishes, if a preparatory course of instruction be arranged, having for its object the training the mind for the study of the science rather than the communicating the knowledge of it. In this preparatory course the order is determined by a consideration of the mind of the pupil; it commences with what is already known to him, and proceeds to the proximate truth; the more easy precedes the more difficult: the *individual* prepares for the *general* truth, the *example* for the *rule.*

It has been objected to the former Edition of this little work that it is not *complete;* it does not profess to be so; it is only a *preparatory* course; the vestibule, not the building. It is proposed that *rules* and practice on rules, should follow these exercises; and this plan has for years been adopted in the school for the use of which the lessons were originally drawn up. The principal alteration which the Second Edition exhibits, is the division of the work into two Parts—the one supplying Directions for the Master, and the other Exercises for the Pupil. This arrangement will materially diminish the manual labour of the teacher, and facilitate his giving instruction to several classes at the same time. It is obvious that a single copy of the larger part will suffice for the Master, and that each of the pupils should be furnished with a copy of the Praxis.

C. MAYO.

Cheam, Jan. 1, 1835.

LESSONS ON NUMBER.

INTRODUCTION.

ON NUMBER.

Idea of Number—Counting.

THE aim of this Lesson is to lead the pupils to understand the meaning of words, expressive o numbers, as "two," "three," "ten," &c.; namely, that they imply a collection of so many ones. To do this, any convenient object, such as a ball, stone, book, &c., being placed before the pupils, the *Teacher* says—*One* ball.

Pupils. [Repeat.] *One* ball.

T. Show me any other object in the room of which you may say *One*.

P. One ceiling, floor, &c.

T. (Putting a second ball to the former, says) *Two* balls.

P. [Repeat.] *Two* balls.

T. Show me any objects of which you may say *Two.*

P. Two doors, &c.

T. (Adding a third ball to the former two, says) *Three* balls.

P. Three balls.

T. Name any objects of which you may say *Three.*

P. Three chairs, &c.

Adding successively a ball to the former number, the teacher each time requires the pupils to point out some objects to which the appellation, four, five, six, &c., is applicable. It must be left to his discretion to determine where to stop. One child will be embarrassed by having ten or twenty objects before him, whereas another will, at one glance, ascertain their number. In the one the power of perception must be developed by a slow and gradual process; in the other, it will rapidly strengthen as larger collections of objects are successively presented.

In order however to lead to the accurate conception and correct expression of number, it is desirable at this step to put before the pupils promiscuous objects, requiring them to ascertain their total number, as well as the number of those among them, which are of the same kind.

Thus, for instance; putting 3 balls, 4 books, 5 slates, &c., before them, the teacher says—

T. What must you do to ascertain how many objects there are here?

P. We must *count* them.

T. Count them.

P. The.e are twelve objects.

T. Twelve, then, is the *number* of objects before you; are they all of the same kind?

P. No; there are balls, books, and slates.

T. Ascertain the number of each class of objects.

P. The number of balls is three; the number of books four; and the number of slates five.

Exercises of this kind may be much and interestingly diversified, especially for very young children, or for those whose perception is slow, by directing their attention to the number of the various parts of which one object is composed, or the number of plants, trees, birds, fishes, shells, minerals, &c., which they know.

The teacher is referred to "Exercises on Lessons on Number."

CHAPTER I.—ADDITION.

§ 1. ADDITION OF UNITS.

LESSON I. *To add One.*

THE first and simplest lesson in Number is evidently that in which the pupil is taught to add one to each number in succession; and then to numbers taken promiscuously. To connect this operation of the mind with the idea the pupil has formed of number from the previous exercises, recourse must be had to the senses, by making use of some visible objects, either the numeral frame or the slate. This being ultimately abandoned, the exercise will become purely intellectual.

In the following exercises the slate is supposed to be used.

T. [Drawing one small line upon the slate, asks] How many lines are there here? |

P. One.

T. [Draws beneath the former two lines, and asks] How many lines are there here? | |

P. Two.

T. How many are one line more one line?

P. Two lines.

T. How many are one book more one book?—One tree more one tree?—One slate more one slate?

How much are one more one?

How many ones make two?

T. [Drawing three lines beneath the former two, asks] How many lines are here? | | |

P. Three.

T. How many more lines are there in this row than in the one above?

P. One more.

T. How many lines are in the second row?

P. Two.

T. How many lines are two more one?

P. Three lines.

T. How much is two more one?

P. Three.

T. [Draws *four* lines beneath the former three, asking] How many more lines are there in this row than in the one above? | | | |

P. One more.

T. How many are in the third row.

P. Three.

T. How many are three lines more one?

P. Four lines.

T. How much is three more one?

P. Four.

T. [Draws *five* lines beneath the former four; *six, seven* lines, and so on in succession, asking each time questions analogous to the above. These lines will stand arranged thus:—

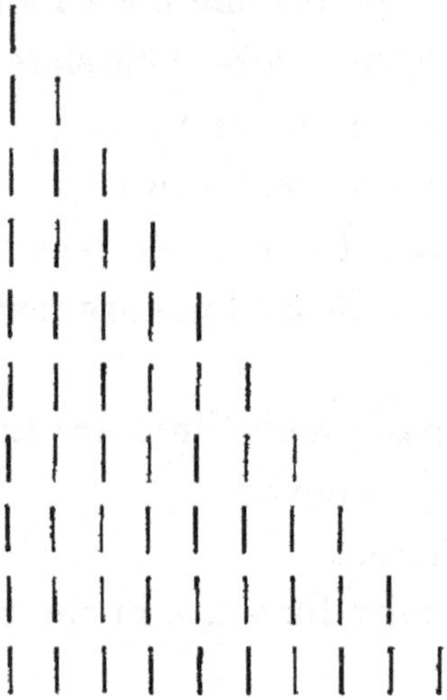

To such a number, as, in the discretion of the teacher, will seem sufficient.

After due repetition, the pupil must be able to proceed as follows: [the lines on the slate being effaced.]

One more one are two;
Two more one are three;
Three more one are four;
Four more one are five;
Five more one are six;
Six more one are seven, &c., &c., &c.

The teacher will do well to let his pupils proceed to the very limit of their distinct conception of number. After this exercise, which the whole class should repeat vivâ voce, each pupil may be required to ask the class one or more questions promiscuously. Thus, How much is nine more one?—Seventeen more one? &c.

It frequently happens that the teacher's class consists of several divisions of pupils, more or less advanced, and the time he can devote to each is, in consequence, limited. Exercises have, therefore, been drawn up corresponding to this and the following lessons, and may be given to the pupils for solution. The answers are, for the present, to be written on their slates in *words*.

Another advantage of these exercises, and that not perhaps the least, is, that they lead each pupil in silence to think for himself, and commit the result of his thoughts to writing;—a process much calculated to sober the mind after the exciting effects of simultaneous learning.

Answers to the Exercises.

Lesson I. *To add* One.

N.B. The answers to the first six questions must be inspected by the teacher.

Ans. 7.	15	*Ans.* 13.	74	*Ans.* 19.	101
8.	27	14.	79	20.	109
9.	38	15.	81	21.	127
10.	49	16.	87	22.	149
11.	56	17.	94	23.	155
12.	63	18.	97	24.	187

LESSON II. *To add* TWO.

T. [Drawing one line on the slate, and next to it two lines more, thus] asks

How many lines are there in all?

P. Three lines.

T. [Drawing two lines, and next to them two lines more, three lines, four lines, &c., adding each time two lines, as below,] asks

How many lines are one more two? two more two? three more two lines? &c., &c.

P. Two lines more two lines are four lines;
Three lines more two lines are five lines;
Four lines more two lines are six lines;
&c., &c.

After due repetition of the above, the drawing of lines on the slate should be abandoned, if possible. It is far more improving to render the operation purely intellectual, than to continue deriving assistance from the external senses.

To add two, is, evidently, to add one, and then another one in succession. If, then, Lesson I. be perfectly known, there will be little difficulty in leading the pupil to increase any number by two at one step. The mode made use of to this end, is shown in the following questions:—

Q. How many ones make two?

A. One more one; that is, two ones.

Q. How many ones, then, make one more two?

A. One more one, more one; that is, three.

Q. How much is two more two?

A. Two more one, more one, or four.

Q. How much is three more two?

A. Three more one, more one, or five.

Q. How much is four more two? five more two? six more two? &c.

The pupils must, after due repetition of such exercises, be able to proceed readily as follows:—

One more two are three;
Two more two are four;
Three more two are five;
Four more two are six;
Five more two are seven; &c., &c.

For exercise, it is recommended to require the pupils to count by twos, as they have learned to count by ones; that is, the teacher begins at any number he pleases, and calls upon the pupils to add two, and to the answer two again, and so on in succession, as far as they are able, or as may be sufficient for practice.

Answers to the Exercises. Lesson II.

Ans. 1.	19	*Ans.* 5.	63	*Ans.* 9.	103
2.	27	6.	71	10.	121
3.	49	7.	82	11.	151
4.	58	8.	91	12.	201

Ans. 13. 1, 3, 5, 7, 9, 11, 13, 15, 17, 19, 21, 23, &c.

Ans. 14. 2, 4, 6, 8, 10, 12, 14, 16, 18, 20, 22, 24, &c.

Lesson III. *To add* Three.

Before entering upon any new lesson, the teacher must be quite certain that the preceding is firmly fixed in the minds of the pupils. On this supposition, the present lesson will require but little effort on their part. They are able to add *two,* also to add *one,* and are now required to add *three;* that is, *two* more *one.*

Teacher. You have learnt to add *two;* if, now, instead of two, you were required to add *three,* what would you do?

Pupils. First add two, and then one more.

T. How much is one more three?

P. One more two, more one, or four.

T. How much is two more three?

P. Two more two, more one, or five.

T. How much is three more three?

P. Three more two, more one, or six.

T. How much is four more three? five more three? six more three? &c., &c.

After due repetition, the pupil must be able to proceed readily thus:—

One more three are four;
Two more three are five;
Three more three are six;
Four more three are seven; &c., &c.

The teacher may now call upon the pupils to give questions promiscuously to the class.

Pupil. How much is 17 more 3?

Class. Twenty.

P. How much is 59 more 3?

C. Sixty-two.

&c., &c.

The whole class then begin to count by adding three in succession; thus:—

C. Three, six, nine, twelve, fifteen, &c. Again,

C. One, four, seven, ten, thirteen, &c. And again,

C. Two, five, eight, eleven, fourteen, &c.

As a repetition and preparation for the next lesson, the teacher may combine the previous two lessons with the last. Thus:—

T. How much is one more three, more one?
... ... two more three, more one?
... ... four more three, more one?
... ... five more three, more one?
&c., &c.

Again,

T. How much is one more two, more two?
... ... two more two, more two?
... ... three more two, more two?
&c., &c.

And again,

T. How much is one more two, more three?
... ... two more two, more three?
... ... three more three, more three?
&c.

Then, promiscuously:—

T. How much is five more three, more two?
... ... seventeen more two, more two?
... ... thirty-nine more two, more three?
... ... forty-five more one, more two, more three?

It will much tend to enliven the pupils, if the teacher permit them to give questions, similar to the above, to the class. For instance, one pupil asks,

How much is forty-three more three, more two?

Those who have found the answer, hold up one hand. The pupil who gave the question chooses among them whom he pleases to answer it; and it is he, likewise, who is to say whether the answer be right or wrong.

Answers to the Exercises. Lesson III.

Ans. 1.	17	*Ans.* 4.	91	*Ans.* 7.	60
2.	32	5.	78	8.	102
3.	40	6.	101	9.	121
				10.	152

Ans. 11. 4, 7, 10, 13, 16, 19, &c.
12. 5, 8, 11, 14, 17, 20, &c.
13. 6, 9, 12, 15, 18, 21, &c.

Ans. 14.	10	*Ans.* 16.	41	*Ans.* 18.	92
15.	23	17.	62	19.	104
				20.	96

Lesson IV. *To add* Four.

This and the following lessons are quite analogous to those preceding; and it is not without reason that the teacher is recommended to observe a *progressive* order in these lessons, always beginning with the simplest form possible. Accordingly, the number four is added to one, then to two, next to three, and so on in succession, as far as the teacher may think proper, within the limit of the pupil's clear conception of number.

The answers, being the numbers 5, 6, 7, &c., in their natural order, are readily found by the pupils, which serves as encouragement; but to arrive at the aim in view, namely, to add four to any number at one step, the teacher requires the pupils to count by fours; that is, he begins with adding four to one; to the answer four again is added, and so on, which gives rise to the following progression:—

1, 5, 9, 13, 17, 21, 25, 29, 33, &c.

Two is then taken as the beginning of the next series; thence the following progression:

2, 6, 10, 14, 18, 22, 26, 30, 34, &c.

Three now begins the series, whence the answers,

3, 7, 11, 15, 19, 23, 27, 31, 35, &c,

And, finally, four is made the beginning from which the answers

4, 8, 12, 16, 20, 24, 28, 32, 36, &c.

It is obvious that by this mode of proceeding, four is added to each of the numbers, 1, 2, 3, 4, &c., and the subject quite exhausted. The pupils must be able to repeat these four series, without hesitating, before they proceed to the next lesson.

Promiscuous questions for practice are left to be given by each pupil to the class.

Answers to the Exercises. Lesson IV.

Ans. 1. 43, 82, 91, 99, 123.
2. 22 *Ans.* 4. 66 *Ans.* 6. 107
3. 43 5. 95
7. 43, 47, 51, 55, 59, 63, &c.
8. 61, 64, 66, 70, 73, 75, 79, 82, 84, 88, 91, 93, 97, 100, 102, 106, &c.

Lesson V. *To add* Five.

The difficulty of adding a number at one step increases with the number itself; but it may be removed in a great measure, by inducing the pupils to use the knowledge already gained. Thus to add five is evidently to add four more one; or to add three more two; or two more two more one; operations already performed by them. The teacher ought therefore to object to the pupils counting by their fingers, but constantly require them to refer to that actually known, and to use it accordingly. The teacher may proceed thus:—

Teacher. You have learnt to add four; if now it were required to add five, what would you do?

Pupil. First add four, and then one more.

T. But if you began with adding three, what then must be done?

P. Add two more.

T. How much then is seventeen more five?

P. Seventeen more four more one, that is twenty-two.

T. Or by adding three first.

P. Seventeen more three more two, which is twenty-two.

T. The answer twenty-two is usually called the *sum* of seventeen and five; and whenever two or more numbers are added together, the answer is called their *sum*. What is the *sum* of thirty-seven and five.

P. Forty-two.

T. What is the sum of one, two, three, four, and five?

P. Fifteen.

We will now learn to add five at one step, without first adding four and then one, or three, and then two.

Here now follow exercises analogous to those in the preceding lessons.

T. How much is one more five?

... ... two more five?

... ... three more five?

... ... four more five?

&c., &c.

After which one is made the beginning of a series, five is added, to the sum five again, and so on. Then two is made the beginning, three next, four next, and finally five, which completes the lesson. The answers, which the pupils must be able to repeat vivâ voce, are respectively as follow—

Class, count 1, 6, 11, 16, 21, 26, 31, 36, 41, &c.

... 2, 7, 12, 17, 22, 27, 32, 37, 42, &c.

... 3, 8, 13, 18, 23, 28, 33, 38, 43, &c.

... 4, 9, 14, 19, 24, 29, 34, 39, 44, &c.

... 5, 10, 15, 20, 25, 30, 35, 40, 45, &c.

This mode of proceeding will soon be remarked by the pupil; he will anticipate the questions of his master; it will awaken his mind to reflection, and lead him to self-activity.

Before entering upon the next lesson, each pupil gives questions to the class, now making use of the word *sum*. Thus:—

P. What is the sum of fifty-eight and five?

&c.

Answers to the Exercises. Lesson V.

Ans. 1.	48	*Ans.* 4.	112.	*Ans.* 7.	114
2.	69	5.	151.	8.	203
3.	99	6.	162.	9.	224

10. 22, 26, 29, 31, 32, 37, 41, 44, 46, 47,
52, 56, 59, 61, 62, 67, 71, 74, 76, 77, &c.

Lesson VI. *To add* Six.

The mode of proceeding being quite in accordance with that detailed in the foregoing lessons, it will be sufficient to give a general outline of the exercises which follow in this paragraph.

1st, *Teacher.* How much is one more six?
... ... two more six?
... ... three more six?
&c. &c.

2d, *Class,* count 1, 7, 13, 19, 25, 31, &c.
... ... 2, 8, 14, 20, 26, 32, &c.
... ... 3, 9, 15, 21, 27, 33, &c.
... ... 4, 10, 16, 22, 28, 34, &c.
... ... 5, 11, 17, 23, 29, 35, &c.
Finally, ... 6, 12, 18, 24, 30, 36, &c.

3d. Promiscuous questions given by each pupil to the Class.

Ex. 1. *Pupil.* How much is fifty-nine more six?

Ex. 2. *P.* Add seventy-three and six.

Ex. 3. *P.* What is the sum of thirty-seven and six.

Ex. 4. *P.* What is the sum of eight, six, five, and four.

N. B.—The pupils should be required to use the words, "*more,*" "*add,*" "*sum,*" in order to form correct notions of their signification.

Answers to the Exercises. Lesson VI.

Ans.	1.	53	*Ans.*	6.	76	*Ans.*	11.	89
	2.	21		7.	126		12.	93
	3.	110		8.	56		13.	110
	4.	118		9.	64		14.	116
	5.	147		10.	78		15.	255

Lesson VII. *To add* Seven.

Outline of the various Exercises of this Lesson.

1st. *Teacher.* How much is one more seven?
... ... two more seven?
... ... three more seven?
&c., &c.

2d. *Class,* count 1, 8, 15, 22, 29, 36, 43, 50, &c.
... ... 2, 9, 16, 23, 30, 37, 44, 51, &c.
... ... 3, 10, 17, 24, 31, 38, 45, 52, &c.
... ... 4, 11, 18, 25, 32, 39, 46, 53, &c.
... ... 5, 12, 19, 26, 33, 40, 47, 54, &c.
... ... 6, 13, 20, 27, 34, 41, 48, 55, &c.
Finally, ... 7, 14, 21, 28, 35, 42, 49, 56, &c.

3d. Promiscuous questions given by the pupils:

Ex. 1. *Pupil.* How much is forty-three more seven?

Ex. 2. *P.* Increase seventy-seven by seven.

Ex. 3. *P.* Add eighty-nine and seven.

Ex. 4. *P.* Find the sum of ninety-eight and seven.

Ex. 5. *P.* Find the sum of one, two, three, four, five, six, and seven.

Answers to the Exercises. Lesson VII.

Ans. 1.	67	*Ans.* 6.	83	*Ans.* 11.	66
2.	77	7.	90	12.	132
3.	111	8.	114	13.	127
4.	124	9.	122	14.	201
5.	129	10.	201	15.	326

Lesson VIII. *To add* Eight.

Outline of this Lesson.

1st. How much is one more eight?
... ... two more eight?
... ... three more eight?

2d. *Class*, count 1, 9, 17, 25, 33, 41, &c.
... ... 2, 10, 18, 26, 34, 42, &c.
... ... 3, 11, 19, 27, 35, 43, &c.
... ... 4, 12, 20, 28, 36, 44, &c.
... ... 5, 13, 21, 29, 37, 45, &c.
... ... 6, 14, 22, 30, 38, 46, &c.
... ... 7, 15, 23, 31, 39, 47, &c.
Finally, ... 8, 16, 24, 32, 40, 48, &c.

3d. Promiscuous questions given by the pupils:

Ex. 1. How much is forty-five more eight?

Ex. 2. Increase seventy-eight by eight.

Ex. 3. Add eighty-seven and eight.

Ex. 4. Find the sum of ninety-six and eight.

Ex. 5. Find the sum of two, seven, five, and eight.

Answers to the Exercises. Lesson VIII.

Ans. 1.	83	*Ans.* 6.	131	*Ans.* 11.	65
2.	109	7.	128	12.	83
3.	124	8.	131	13.	89
4.	143	9.	204	14,	122
5.	251	10.	335	15.	134

Lesson IX. *To add* Nine.

Outline of this Lesson.

1st. *Teacher.* How much is one more nine?
... ... two more nine?
... ... three more nine?
&c., &c.

2d. *Class*, count 1, 10, 19, 28, 37, &c.
... ... 2, 11, 20, 29, 38, &c.
... ... 3, 12, 21, 30, 39, &c.
... ... 4, 13, 22, 31, 40, &c.
... ... 5, 14, 23, 32, 41, &c.
... ... 6, 15, 24, 33, 42, &c.
... ... 7, 16, 25, 34, 43, &c.
... ... 8, 17, 26, 35, 44, &c.
Finally, ... 9, 18, 27, 36, 45, &c.

3d. Promiscuous questions given by the pupils:

Ex. 1. *Pupil.* How much is sixty-three more nine?

Ex. 2. *P.* Increase forty-seven by nine.

Ex. 3. *P.* Add seventy-eight and nine.

Ex. 4. *P.* What is the sum of ninety-two and nine?

Ex. 5. *P.* Find the sum of thirteen, nine, seven, and six.

Answers to the Exercises. Lesson IX.

Ans.	1.	50	*Ans.*	6.	72	*Ans.*	11.	193
	2.	100		7.	84		12.	224
	3.	99		8.	129		13.	243
	4.	116		9.	101		14.	262
	5.	174		10.	138		15.	431

Lesson X.

Decomposition of Numbers into Tens and Units.

The decimal system of numeration being that generally adopted, it is of importance to lead the pupils, as far as their tender years allow, to the perception of its advantages: and this may be done at the present stage.

Teacher. Count as far as ten.

Pupils. [Counting,] One, two, ... ten.

T. The number next above ten is eleven—that is, ten more ——?

P. Ten more one.

T. The number following is twelve; that is—

P. Ten more two.

T. Continue counting, thirteen is—

P. Thirteen is ten more three;
Fourteen is ten more four;
Fifteen is ten more five;
Sixteen is ten more six;
Seventeen is ten more seven;
Eighteen is ten more eight;
Nineteen is ten more nine;
Twenty is ten more ten;
Twenty-one is ten more eleven.

T. You said before, that eleven was ten more one; what, then, will you say that twenty-one is?

P. Ten more ten, more one.

T. That is, how many tens?

P. Two tens more one.

T. Continue, twenty-two is—

P. Twenty-two is two tens more two.

T. Two what?

P. Two ones.

T. Proceed: twenty-three is—

P. Twenty-three is two tens more three ones;
Twenty-four is two tens more four ones,
&c., &c.
Twenty-nine is two tens more nine ones.

T. Instead of saying *ones*, it is usual to say *units.* Proceed.

P. Thirty is two tens more ten units.

T. That is?

P. Three tens.
Thirty-one is three tens more one unit,
&c., &c.
Forty is three tens more ten units; that is four tens.

The pupils are required to continue, in a similar manner, as far as the teacher may judge necessary.

T. Hence it is easy to say of how many tens and units a number consists.

Thus, eighty-eight consists of how many units?

P. Of eighty-eight units.

T. And of how many tens?

P. Of eight tens and eight units.

T. By which part of your body are you apt to count?

P. By our fingers.

T. And how many fingers have we?

P. Ten.

T. Whence, then, do you think originates our counting thus by tens?

P. Probably from the circumstance of our having ten fingers.

T. How many tens are in eleven more ten?

P. Two tens more one unit.

T. How many units is that?

P. Twenty-one.

T. Continue adding ten to twelve, thirteen, &c.

P. Twelve more ten is two tens more two units, or twenty-two;

Thirteen more ten is two tens more three, or twenty-three;

Fourteen more ten is two tens more four, or twenty-four;

Fifteen, &c.

T. Hence, if ten is to be added to a number, what must be done?

P. Only increase the number of tens by one ten.

T. Give an example.

P. Thirty-eight more ten, is three tens more eight units, more *one* ten; that is, four tens more eight units, or forty-eight.

It will be convenient, before proceeding further, to teach the pupils how to represent numbers by signs, not that they are *essential* to the succeeding lessons, which are intended as practice in *mental* arithmetic; but because the *written* exercises would, without the aid of signs, become cumbrous. In a school, then, where the teacher's time is divided among several classes, more or less advanced, the necessity of the exercises will be felt; and for such it is recommended to have recourse to written arithmetic: whereas in a family of a few children, no such necessity existing, it will be better to defer it, and continue the exercises purely *mental*, until the pupils' minds, from the following lessons, are more developed.

How to represent Units by Signs.

I.—UNITS.

Teacher. We will now learn to represent numbers by signs [writing the following on the large school-slate:]

|	1.
| |	2.
| | |	3.
| | | |	4.
| | | | |	5.
| | | | | |	6.
| | | | | | |	7.
| | | | | | | |	8.
| | | | | | | | |	9.

[Pointing to the lines, and their respective representatives in number, says, and pupils repeat,] One, two, three, &c. These signs are called *figures;* imitate them on your own slates.

T. What numbers do these figures represent?

Pupils. Ones, or units.

T. Hence the figure 2 is equal in value to which two figures added together?

P. To 1 more 1.

T. I will teach you now another sign, generally used to denote the word "*more;*" it is this +. If, then, we wish to write down what we have just said, we would, instead of writing thus,

2 is equal in value to 1 more 1,

write, 2 is equal in value to 1 + 1.

For the words, "is equal in value," or "are equal in value," it has been agreed to substitute the sign =; so that

2 is equal in value to 1 more 1,

is written, 2 = 1 + 1.

What advantage has this sort of short-hand writing?

P. It saves time and room.

T. Let one of you come here to the slate, and, by means of these signs, write which three numbers the figure 3 is equal to.

P. [*Writing.*] 3=1+1+1.

T. And which two figures is 3 equal to?

P. [*Writing.*] 3=2+1.

T. Proceed in a similar manner with the figure 4.

P. [*Writing.*] 4=1+1+1+1.

4=2+1+1.

4=2+2.

4=3+1.

T. Let, now, every one of you proceed in the same way with the remaining figures, 5, 6, 7, 8, and 9, on your own slates.

II.—Tens.

T. What is the number next above nine?

P. Ten.

T. Represent ten by the addition of two of the figures you have learnt.

P. $9+1$; $8+2$; $7+3$, &c.

T. How many ones, or units, are there in ten?

P. Ten units.

T. And how many tens?

P. One ten.

T. Which of the figures represents one?

P. 1.

T. If, now, we wish to represent one ten, will it be sufficient to write 1?

P. No; because we could not tell whether one was meant or ten.

T. In order to raise this difficulty, I will place this sign, 0, called zero, next to the 1, showing that the 1 is meant for 1 ten; thus, 10. Tell me, now, which place 0 occupies with respect to the 1.

P. It stands behind the 1; it stands before the 1.

T. Some of you said, "behind," others, "before;" which is right?

P. "Behind;" because we read and write from the left to the right.

T. So we do with the letters of the alphabet; but are you sure it is the same when signs are used instead of letters? You mentioned "left" and "right;" which of those places does 0 occupy in respect to 1?

P. It stands to the *right* of it.

T. Well; tell me, now, how many units there are in ten.

P. Ten units.

T. And how many tens?

P. One ten.

T. So that if we suppose these ten units collected, there would be *one* ten, and no more units. Which, now, of the figures 10, indicates *one* ten, and which no units?

P. The one in 10 indicates one ten, and the zero no units.

T. And you have before remarked that the figure zero stands where?

P. To the right of one.

T. Hence we shall for the future agree, that, of two figures, that which stands to the right of the other indicates what?

P. Units.

T. And what to the left?

P. Ten.

T. How am I to represent two tens, or twenty?

P. Write the figure two, and to the right of it zero.

T. Let one of you come here and write three tens, or thirty; four tens, or forty; fifty, sixty, &c., ninety.

T. How many tens are there in one hundred?

P. Ten tens.

T. You have learnt to write ten; we have agreed upon the place which the tens should occupy;—try and express one hundred.

It must be left to the pupils to represent that number correctly; when done, the teacher proceeds:—

T. How many figures have you used to represent one hundred?

P. Three figures.

T. How many units are there in one hundred?

P. One hundred units.

T. How many tens?

P. Ten tens.

T. And how many hundreds?

P. *One* hundred.

T. So that if we suppose these hundred units collected into tens, there would be how many of them?

P. Ten, and no units.

T. And again; if we suppose these tens collected into hundreds, there would be ——

P. One of them, and no tens.

T. Which, now, of the figures 100, indicates no units, which no tens, and which one hundred?

P. The last zero to the right indicates no units; the next, to the left of the former, indicates no tens; and the one to the left of this indicates one hundred.

T. Hence we will for the future agree, that, of three figures, that which stands to the right of the two others, shall indicate ——

P. Units.

T. And that which stands next to the left of the former, shall indicate ——

P. Tens.

T. Finally, that which stands to the left of this, shall indicate ——

P. Hundreds.

T. Now represent on your slates two hundred, three hundred, four hundred, &c.

From this it will be readily perceived how the pupils may be taught to represent thousands, tens of thousands, &c.; the teacher remembering that the mind of the pupil will be strengthened by pursuing these exercises so long as he conceives with clearness, but no longer.

As before with units so now the pupils are required to proceed with tens. Thus:—

$$20=10+10.$$
$$30=10+10+10=20+10.$$
$$40=10+10+10+10=20+10+10.$$
$$=20+20=30+10.$$

&c., &c.

III.—Tens and Units.

Teacher. Mention the numbers which are between ten and twenty.

Pupils. Eleven, twelve, &c. .... nineteen.

T. Bearing in mind what places it was agreed that units and tens should occupy, try to represent the number eleven.

The pupils must be left to represent that number correctly. The right or the wrong the class must decide; for that purpose it is best to call each in turn to the school slate. Should eleven be written thus 101, the teacher asks—

T. We agreed that the last figure to the right shall represent units; the next to the left of it tens; the third hundreds, &c. Does this number represent eleven, &c., &c., until the correct answer 11 is found.

T. Here then are two ones, the one to the right of the other indicates ——

P. One unit.

T. That figure is then said to stand in the *units'* *place;* the other 1 to the left of the former indicates ——

P. One ten.

T. It is said to stand in the *tens'* place; now represent twelve, thirteen, &c.

In a manner quite analogous to the above, the pupils are led to represent the numbers, twenty-one, twenty-two, &c., &c., which will not be attended with any difficulty, if the preceding has been understood. The pupils must discover themselves that all numbers between 10 and 100 are properly represented by 2 figures, those between 100 and 1000 by 3 figures, and so on. For practice, it is recommended to require the pupils to write in figures the questions and their answers according to the following model: thus, beginning at—

Lesson I. Quest. 7. The pupils write on their slates—

$$14+1=15.$$
$$26+1=27.$$
$$37+1=38.$$
&c.

Lesson II. Quest. 1.

$$17+2=19.$$
$$25+2=27.$$
$$47+2=49.$$
&c.

Lesson III. Quest. 1.

$$14+3=17.$$
$$29+3=32.$$
$$37+3=40.$$
&c., &c.

And so on with the lessons which follow:—

Thus, Lesson IX. Quest. 1.

$$5+9+4+9+3+9+2+9=50.$$

The mental exercises are now resumed; but those which the pupils have to perform, as practice on their slates, are now written in figures instead of words.

§ 2. ADDITION OF TENS.

LESSON I. *To add* Ten.

Teacher. How much is one more one?
Pupils. Two.
T. How much is one ten more one ten?
P. Two tens.
T. That is —
P. Twenty units.
T. How much is two more one?
P. Three.
T. How much is two tens more one ten?
P. Three tens; that is, thirty.

T. How much is three more one?

P. Four.

T. How much is three tens more one ten?

P. Four tens, or forty.

The teacher thus continues, first, by asking how much is four more one, then four tens more one ten; five more one, then five tens more one ten; and so on, until the pupils are able to proceed readily. Thus:—

Pupils count: ten more ten are twenty;
twenty more ten are thirty;
thirty more ten are forty;
&c., &c.

On comparing this lesson with Lesson I, Addition of Units, it will be found to correspond with it in every step; with this difference, that there is an addition of tens instead of units. The lessons which follow in this paragraph, namely, those which refer to the addition of two tens or twenty, three tens or thirty, &c., precisely follow the same course as those which had for their object to add two, three, &c.; and treated in this manner, present but little or no difficulty. From this consideration, it is thought sufficient to give an outline of one or two lessons on this subject, and leave it to the teacher to supply those which remain.

Answers to the Exercises.

Lesson I. *To add* Ten.

The teacher, on examining the slates, must find

the questions and their answers drawn up as follows:—

Ans. 1.	40+10=50.	*Ans.* 6.	150+10=160.
2.	70+10=80.	7.	160+10=170.
3.	80+10=90.	8.	190+10=200.
4.	90+10=100.	9.	280+10=290.
5.	120+10=130.	10.	360+10=370.

LESSON II. *To add* Twenty.

Teacher. How much is one more two?

Pupils. Three.

T. How much is one ten more two tens?

P. Three tens, that is thirty.

T. How much is two more two?

P. Four.

T. How much is two tens more two tens?

P. Four tens, that is forty.

The teacher continues questioning, in a similar manner, until the pupils are able to proceed readily. Thus:—

Class count: ten more twenty are thirty;
twenty more twenty are forty;
thirty more twenty are fifty;
&c., &c., &c.

Then follow promiscuous questions, given as usual by the pupils to the class. Thus:—

Pupil 1. How much is ninety more twenty?
2. Find the sum of 270 more 20.
3. Add 380 and 20.

Answers to the Exercises. LESSON II.

Model how the exercises are to be written by the pupils:—

Ans. 1. 50+20=70

2. 90+20+10=120.

Ans. 3.	130	*Ans.* 7.	180
4.	150	8.	210
5.	210	9.	280
6.	120	10.	390

LESSON III. *To add* Thirty.

Teacher. How much is one more three?
How much is one ten more three tens?
How much is ten more thirty?

The questions being answered satisfactorily, the teacher proceeds:—

T. How much is two more three?
How much is two tens more three tens?
How much is twenty more thirty?

And so on with thirty more thirty, forty more thirty, &c.; and it will be necessary to begin always with the addition of units, and then transfer the same to the corresponding tens. The pupils must, at the end of this lesson, be able to proceed. Thus:—

Class count: ten more thirty are forty;
twenty more thirty are fifty;
thirty more thirty are sixty;
&c., &c., &c.

Then follow promiscuous questions given by the pupils:—

Pupils 1. Add 90 to 30.

2. Find the sum of 50, 30, and 20.

3. Increase 180 by 30.

Answers to the Exercises. Lesson III.

Model. *Ans.* 1. 70+30+20+10=130.

Ans. 2.	200	*Ans.* 6.	180	*Ans.* 10.	500
3.	270	7.	340	11.	530
4.	300	8.	130	12.	650
5.	140	9.	370		

LESSON IV. *To add* Forty.

Answers to the Exercises. Lesson IV.

Ans. 1.	100	*Ans* 6.	310
2.	210	7.	390
3.	280	8.	590
4.	470	9.	670
5.	570	10.	760

LESSON V. *To add* Fifty.

Answers to the Exercises. Lesson V.

Ans. 1.	150	*Ans.* 6.	250
2.	230	7.	450
3.	300	8.	600
4.	380	9.	710
5.	420	10.	1090

Lesson VI. *To add* Sixty.

Answers to the Exercises. Lesson VI.

Ans. 1.	210	*Ans.* 6.	540
2.	250	7.	550
3.	350	8.	720
4.	330	9.	810
5.	410	10.	960

Lesson VII. *To add* Seventy.

Answers to the Exercises. Lesson VII.

Ans. 1.	280	*Ans.* 4.	390
2.	400	5.	640
3.	390	6.	1090

Lesson VIII. *To add* Eighty.

Answers to the Exercises. Lesson VIII.

Ans. 1.	360	*Ans.* 4.	730
2.	490	5.	770
3.	580	6.	1190

Lesson IX. *To add* Ninety.

Answers to the Exercises. Lesson IX.

Ans. 1.	440	*Ans.* 4.	600
2.	500	5.	860
3.	460	6.	1210

§ 3. ADDITION OF TENS AND UNITS.

LESSON I. *To add* 10, 11, 12...........19.

Thus prepared, the pupils must now begin to add numbers consisting of tens and units, commencing again with the simplest form; thus:—

Teacher. How much is 1 more 10?
... ... 2 more 10?
... ... 3 more 10?
&c., &c.
... ... 10 more 10?
... ... 11 more 10?

Pupils. 10 more 1, more 10; that is, 21.

T. How much is 12 more 10?

P. 10 more 2, more 10; that is, 22.

T. How much is 13 more 10?

P. 10 more 3, more 10; that is, 23.

T. How much is 14 more 10?
... ... 15 more 10?
&c.
... ... 20 more 10?
... ... 21 more 10?

P. 20 more 1, more 10, or 31.

T. How much is 22 more 10?

P. 20 more 2, more 10, or 32.

T. How much is 23 more 10?
... ... 24 more 10?
... .. 25 more 10?
&c.
... ... 31 more 10?

P. 30 more 1, more 10, or 41.

T. How much is 32 more 10?

... ... 33 more 10?

... ... 34 more 10?

&c., &c.

And so on with the numbers which follow. The pupils will soon be able to proceed readily thus:—

Class count: 44 more 10 are 54;

45 more 10 are 55;

&c.

73 more 10 are 83;

&c., &c.

T. You have learnt to add 10 to every number; if, now, it were required to add 11, what would you do?

P. First add 10, and then 1 more.

T. How much is 15 more 11?

P. 15 more 10, more 1, or 26.

T. How much is 39 more 11?

P. 39 more 10, more 1, or 50.

T. And if it were required to add 12, what then would you do?

P. First add 10, and then 2 more.

T. How much is 57 more 12?

P. 57 more 10, more 2, or 69.

T. How much is 85 more 12?

P. 85 more 10, more 2, or 97.

T. If it were required to add 13, what would you do?

P. First add 10, and then 3 more.

T. Add 88 and 13.

P. 88 more 13, are 88 more 10, more 3, or 101.

T. If it were required to add 14, what would you do?

P. First add 10, and then 4 more.

T. How much is 137 more 14?

P. 137 more 10, more 4, or 151.

To add the numbers, 15, 16, 17, 18, and 19, similar questions are asked. The above must be considered as a mere outline of the mode of proceeding, not as being sufficient practice in the addition of these numbers. Each number may be taken for the subject of one lesson; and before the pupils have recourse to the written exercises, verbal solutions must be given of questions referring to that number. The pupils should give questions to the class.

Answers to the Exercises. Lesson I.

Models. $57+10=50+7+10=67.$

$83+11=83+10+1=94.$

$149+12=149+10+2=161.$

$217+19=217+10+9=236.$

	To add 10.		*To add* 11.		*To add* 12.
Ans. 1.	59	*Ans.* 1.	94	*Ans.* 1.	61
2.	68	2.	68	2.	65
3.	86	3.	106	3.	9[illegible]
4.	105	4.	114	4.	98
5.	111	5.	128	5.	105
6.	137	6.	165	6.	117
7.	163	7.	160	7.	170
8.	210	8.	187	8.	247

	To add 13.		*To add* 14.		*To add* 15.
Ans. 1.	61	*Ans.* 1.	51	*Ans.* 1.	79
2.	70	2.	72	2.	90
3.	99	3.	83	3.	102
4.	108	4.	87	4.	111
5.	130	5.	100	5.	132
6.	172	6.	111	6.	138
7.	201	7.	128	7.	153
8.	206	8.	150	8.	157
9.	218	9.	171	9.	166
10.	251	10.	293	10.	193

To add 16.

Ans. 1.	74	*Ans.* 4.	114	*Ans.* 7.	165
2.	91	5.	133	8.	169
3.	105	6.	140	9.	202
				10.	213

To add 17.		*To add* 18.	
Ans. 1.	84	*Ans.* 1.	77
2.	102	2.	94
3.	114	3.	112
4.	153	4.	131
5.	160	5.	155
6.	194	6.	176
7.	212	7.	147
8.	248	8.	204
9.	275	9.	315
10.	306	10.	336

To add 19.

Ans. 1.	77	*Ans.* 5.	146	*Ans.* 9.	236
2.	106	6.	151	10.	278
3.	112	7.	194	11.	327
4.	133	8.	202	12.	408

From the mode hitherto pursued, the teacher will perceive how he should treat the addition of the numbers which follow. We have only to guard him against introducing any ready mode of solving the exercises; as, for instance, that of arranging the numbers according to their local values. These exercises are only valuable in as much as they are solved mentally. The work upon the slate must exhibit the process of the mind: thus, if it be required to add 79 and 48, the pupils proceed mentally, in the following manner:—

79 more 48, are 79 more 40, more 8, which are 119 more 8; that is, 127.

And upon the slate the same has the following form:—

$$79+48=79+40+8=119+8=127.$$

And this mode of showing the work upon the slate it is advisable to continue till the pupils have acquired sufficient practice in numeration; after which it may be abandoned, the pupils simply writing down the answer in the usual manner.

Lesson II. *To add* 20, 21, 22...... 29.

Answers to the Exercises.

Ans.	1.	57	*Ans.* 9.	207	*Ans.*	17.	405
	2.	66	10.	226		18.	412
	3.	98	11.	274		19.	425
	4.	111	12.	296		20.	440
	5.	121	13.	318		21.	449
	6,	138	14.	330		22.	66
	7.	154	15.	363		23.	75
	8.	165	16.	394		24.	84

LESSON III. *To add* 30, 31, 32.........39.

Ans. 1.	88	*Ans.* 8.	271	*Ans.* 15.	562
2.	118	9.	304	16.	93
3.	131	10.	386	17.	99
4.	150	11.	480	18.	102
5.	170	12.	500	19.	108
6.	209	13.	522	20.	114
7.	233	14.	540		

LESSON IV. *To add* 40, 41, 42.........49.

Ans. 1.	129	*Ans.* 6.	365	*Ans.* 11.	178
2.	138	7.	375	12.	186
3.	190	8.	404	13.	193
4.	282	9.	166	14.	130
5.	320	10.	170		

LESSON V. *To add* 50, 51, 52.........59.

Ans. 1.	123	*Ans.* 6.	232	*Ans.* 11.	169
2.	134	7.	241	12.	176
3.	149	8.	292	13.	170
4.	172	9.	376	14.	201
5.	190	10.	433		

Lesson VI. *To add* 60, 61.........69.

Ans. 1.	127	*Ans.* 6.	340	*Ans.* 11.	216
2.	146	7.	384	12.	241
3.	189	8.	414	13.	272
4.	241	9.	453	14.	387
5.	258	10.	496		

Lesson VII. *To add* 70, 71.........79.

Ans. 1.	128	*Ans.* 5.	330	*Ans.* 9.	665
2.	151	6.	464	10.	733
3.	169	7.	514	11.	207
4.	216	8.	521	12.	238

Lesson VIII. *To add* 80, 81.........89.

Ans. 1.	156	*Ans.* 5.	297	*Ans.* 9.	752
2.	178	6.	480	10.	865
3.	227	7.	544	11.	290
4.	261	8.	624	12.	224

Lesson IX. *To add* 90, 91.........99.

Ans. 1.	177	*Ans.* 5.	569	*Ans.* 9.	917
2.	234	6.	632	10.	1042
3.	350	7.	783	11.	264
4.	479	8.	871	12.	256

LESSON X. *Promiscuous Questions.*

Note. The numbers, 21, 41, 61, 81, and 79, are not included in the first five questions.

Ans. 1.	210	*Ans.* 4.	755	*Ans.* 7.	792
2.	355	5.	870	8.	940
3.	555	6.	381	9.	2082
				10.	2492

CHAPTER II.—SUBTRACTION.

Lesson I. *To subtract* 1, 10, *and* 11.

The mode of proceeding in subtraction may be anticipated from the manner in which addition has been treated; according to which, 1 would be subtracted, then 2, next 3, and so on. But since the pupils are, at this stage, acquainted with the distribution of numbers into tens, it will be in no wise unsuitable to avail ourselves of their knowledge, and to apply it to the subtraction of tens simultaneously with that of units. Thus, then, if the pupils once know how to subtract one, they will require but a slight effort to subtract one ten, one hundred, one thousand, &c., and even to subtract one ten more one, that is, 11. Again, if the subtraction of 2 be known, that of 2 tens, 2 hundreds, &c., of 10+2, and of 2 tens+2, may be considered as necessary consequences. A similar plan is to be adopted with regard to the numbers, 3, 30, 300, &c.; 13, 23, and 33; and with the numbers following, as will be shown in the lessons.

Teacher. From 1 take away 1; what remains?

Pupils. Nothing.

T. Instead of "take away," it is usual to say *subtract*; hence, from 1 subtract 1, what remains?

P. Nothing.

T. What word did we use when we had to add 1 and 1?

P. We said 1 more 1.

T. Here, now, we have to take away or to subtract 1; what word will express this shortly?

P. 1 less 1.

T. In your written exercises you substituted a sign for the word *more;* what was it?

P. The sign +.

T. Instead of the word "less," we will use another sign, viz., this, —. Express, now, on your slates the result obtained by taking 1 from 1.

P. $1-1=0$.

T. What is the result called, when two or more numbers are added?

P. Their sum.

T. What would you call the result obtained by taking one number from another number?

P. Their *difference.*

T. From 2 take 1; what remains?

From 2 subtract 1; what remains?

What is the difference between 2 and 1?

From 3 take 1; what remains?

From 3 subtract 1; what remains?

What is the difference between 3 and 1?
How much is 4 less 1?
... ... 5 less 1?
&c., &c.

The result of these exercises must be, that the pupils are able readily to proceed thus:

Class count: 1 less 1 is 0;
2 less 1 is 1;
3 less 1 is 2;
4 less 1 is 3;
&c., &c.
100 less 1 is 99.

To subtract 10.

Teacher. From what numbers can 10 be subtracted?

Pupils. From 10, and those numbers which are more than 10.

T. From 10 take 10; what remains?

P. Nothing.

T. Decompose 11 into tens and units.

P. 11 is 10 more 1.

T. If, then, from 11, 10 be taken away, what must remain?

P. 1.

T. Decompose 12 into tens and units.

P. 12 is 10 more 2.

T. If, then, from 12, 10 be taken away, what remains?

P. 2.

T. [Proceeds in a similar manner with 13, 14, 15, &c.] Decompose 19 into tens and units. If, now, 10 be taken from 19, what remains?

P. 9.

T. Decompose 20 in tens.

P. 20 is 2 tens.

T. If, then, from 20, or 2 tens, 1 ten be taken, what remains?

P. 10.

T. Decompose 21 into tens and units.
Take 10 from 21, what remains?
Decompose 22 into tens and units.
Take 10 from 22, what remains?

The pupils will experience no difficulty in subtracting 10 from the numbers which follow, if the subject be treated in this manner. The result must be, that the class will be able to proceed readily thus:

Class count: 10 less 10 is 0;
11 less 10 is 1;
12 less 10 is 2;
13 less 10 is 3; &c., &c.
78 less 10 is 68;
100 less 10 is 90; &c., &c.

To subtract 11.

Teacher. You have learnt to subtract 1, and likewise 10; can you tell what other number you are now able to subtract?

Pupils. 11.

T. Why?

P. Because 11 is 10 more 1.

T. From which numbers can 11 be subtracted?

P. From 11, and those numbers which are more than 11.

T. If, then, it were required to take 11 from any of these numbers, what would you do?

P. First take away 10 and then 1.

T. Subtract 11 from 47.

P. 47 less 11 is 47 less 10, less 1; 47 less 10 is 47; 37 less 1 is 36, the answer.

The pupils are now required to give questions to the class; after which the exercises, Part II, are taken up for further practice.

Answers to the Exercises. Lesson I.

Model. 100−1=99. 57−10=47.
304−11=304−10−1=294−1=293.

Ans. 1. 99, 98, 97, &c.
2. 203, 193, 183, 173, &c.
?. 336, 325, 314, 303, 292, &c.

Ans. 4.	140	*Ans.* 6.	879	*Ans.* 8.	990
5.	769	7.	893	9.	1034
				10.	1293

Lesson II.

To subtract 2, 12, 20, *and* 22.

In this lesson the pupils have actually to learn to subtract 2 from the units, and from 10 and 11; this being known, they must be led to refer all further subtractions of 2, 12, 20, and 22 to these results.

Teacher. If 2 were to be taken from any number what would you do?

Pupils. First take away 1, and then 1 again.

T. Put on your slates the numbers 2, 3......11; from each take away 2, and commit the answers to memory. This done,

Class counts: 2 less 2 is 0;
3 less 2 is 1;
4 less 2 is 2;
&c.
11 less 2 is 9.

T. Do you think it necessary to learn by heart the answers obtained by subtracting 2 from 12, 13, 14, &c., and all other numbers?

P. No; for 12 is 10 more 2;
13 is 10 more 3;
14 is 10 more 4; &c.

and we know what remains if 2 be taken from 2;

hence we know too what remains if 2 be taken from 10 more 2, that is from 12; and so on with the other numbers.

T. Take 2 from 21; how will you proceed in this instance?

P. 21 is twenty more 1, or 10 more 11; and 2 taken from 11 is 9; therefore 2 taken from 21 is 19.

T. Name another number which would cause you to proceed in a similar manner.

P. 31, 41, 51, &c.

T. Give questions to the class.

To subtract 12, 20, 21, *and* 22.

Teacher. Since you know now how to subtract 2, can you tell what other numbers you are now able to subtract?

Pupils. 10 more 2 or 12.

T. Besides 12 there are other numbers which you will be able to subtract very easily; which are they?

P. 2 tens or 20; 20 more 1 or 21; and 20 more 2 or 22.

T. How much is 57 less 12?

P. 57 less 12 is 57 less 10 less 2;
57 less 10 is 47; 47 less 2 is 45;
therefore 57 less 12 is 45.

T. How much is 89 less 20?

P. 89 less 20 is 80 more 9 less 20;
or, 8 tens less 2 tens, more 9, which are 6 tens more 9, or 69.

T. How much is 90 less 21?

P. 90 less 21 is 90 less 20 less 1;
90 less 20 is 70; 70 less 1 is 69, the answer.

T. How much is 71 less 22?

P. 71 less 22 is 71 less 20 less 2;
71 less 20 is 51; 51 less 2 is 49, the answer.

The pupils are required to give similar solutions to the questions. The teacher is recommended to proceed slowly, and to dwell upon each subject until the pupils have acquired sufficient facility.

Answers to the Exercises. Lesson II.

Ans. 1. 98, 96, 94, 92, 80, &c.
2. 188, 176, 164, 152, 140, 128, &c.
3. 387, 367, 347, 327, 307, &c.
4. 359, 338, 317, 296, 275, 254, 233, 212, 191, 170, 149, &c.
5. 329, 307, 285, 263, 241, 219, 197, 175, 153, 131, 109, 87, 65, 43, 21.

Ans.	6.	49	*Ans.*	11.	159	*Ans.*	16.	549
	7.	75		12.	178		17.	595
	8.	73		13.	325		18.	680
	9.	83		14.	395		19.	751
	10.	97		15.	487		20.	839
							21.	879

Lesson III.

To subtract 3, 13, 23, 30, 32, and 33.

What the pupils have actually to learn is, to subtract 3 from 3, 4, 5, 6, 7, 8, 9, 10, 11, and 12. This being known, but a slight effort on their part will enable them to subtract 3 from the numbers above 12.

Again, they know how to subtract 10; they likewise know how to subtract 3; they will experience no difficulty in subtracting 10+3, or 13.

The same remark applies to the subtraction of 20+3, or 23.

Also, if the pupils know how to subtract 3, they must, for the same reason, be able to subtract 3 tens, or 30; and, finally, combining this and the previous lessons with the subtraction of 30, to apply the same reasoning to subtract 30+2, and 30+3.

The mode to which the teacher has to resort, in order to obtain from his pupils the above result, has, it is thought, been sufficiently shown in the previous lessons. There is nothing to prevent applying the same reasoning to the subtraction of 300, 303, 313, 323, 330, 332, and 333, and even thousands. The questions, however, have only been adapted to tens, from the consideration that these exercises are intended for children, whose age will justify us in being content to receive a ready and correct answer to a question similar to the following: "What is the difference between 705 and 33?"

Answers to the Exercises. Lesson III.

Ans. 1. 84, 81, 78, 75, 72, 69, 66, 63, 60, 57, 54, &c.
2. 92, 79, 66, 53, 40, 27, 14, 1.
3. 148, 125, 102, 79, 56, 33, 10.
4. 171, 141, 111, 81, 51, 21.
5. 239, 207, 175, 143, 111, 79, 47, 15.
6. 309, 276, 243, 210, 177, 144, 111, 78, 45, 12.

Ans. 7.	155	*Ans.* 11.	385	*Ans.* 15.	578
8.	158	12.	447	16.	668
9.	129	13.	487	17.	771
10.	275	14.	539	18.	772

Lesson IV.

To subtract 4, 14, 24, 34, 40, 41, 42, 43, and 44.

What the pupils have actually to learn in this lesson is, to subtract 4 from 4, 5, 6, &c....13; this known, the pupils must be *left to discover* that,

1. They are now able to subtract 4 from the succeeding numbers 14, 15, &c.
2. Also 10+4 from 14, 15, &c.
3. Also 20+4 from 24, 25, &c.

4. Likewise 30+4 from 34, &c.

5. Again, 40 from 40, 41, &c.

6. Similarly, 40+1, 40+2, 40+3, and, finally, 40+4 from 41, &c.; 42, &c.; 43, &c.; and 44, &c.

Let it be understood that the pupils must make the discovery themselves, otherwise they will not feel interested. It is also necessary to dwell upon each subject, to give questions, and let each of the pupils give questions, which are to be solved mentally, before the exercises, Part II, are taken up.

Answers to the Exercises. Lesson IV.

Ans. 1. 93, 89, 85, 81, 77, 73, 69, 65, 61, 57, 53, &c.

2. 97, 83, 69, 55, 41, 27, 13.

3. 199, 175, 151, 127, 103, 79, 55, 31, 7.

4. 169, 135, 101, 67, 32.

5. 247, 207, 167, 127, 87, 47, 7.

6. 329, 288, 247, 206, 165, 124, 83, 42, 1.

7. 359, 317, 275, 233, 191, 149, 107, 65, 23.

8. 489, 446, 403, 360, 317, 274, 231, 188, 145, 102, 59, 16.

9. 559, 515, 471, 427, 383, 339, 295, 251, 207, 163, 119, 75, 31.

By subtracting the lesser number from the greater.

Ans.	10. 57	*Ans.*	14.	89	*Ans.*	18.	529
	11. 118		15.	159		19.	599
	12. 49		16.	189		20.	668
	13. 67		17.	279		21.	779

Lesson V.

To subtract 5, 15, 25, 35, 45, 50, 51, 52, 53, 54, and 55.

In this lesson, again, the pupils have only to learn to subtract 5 from 5, 6, 7, &c.......14, and then to apply the rest to these results; and since, from the mode of proceeding in the previous lessons, the pupils must have become aware, by this time, of the sort of inquiries to which the subject necessarily gives rise, the teacher ought to generalise his questions more; thus:

Teacher. When in our last lesson you learnt to subtract 4, from what numbers did you find it sufficient to know how to subtract 4?

Pupils. From 4, 5, 6, 7, 8, 9, 10, 11, 12, and 13.

T. When you knew this, what followed?

P. We then could take away 4 very easily from every other number.

T. Can you tell from which numbers you must learn to subtract 5, so as to be able to subtract 5 readily from every number?

P. From 5, 6, 7, 8, 9, 10, 11, 12, 13, and 14.

T. If you know this, can you then subtract 5 from every number?

P. Yes; for 15 is 10+5; and if we know how to subtract 5 from 5, we likewise know how to subtract 5 from 10+5, or from 20+5, 30+5, &c, And if we know how to subtract 5 from 6, we likewise know how to subtract 5 from 10+6, 20+6, 30+6, &c., and so on with the other numbers.

T. Then write on your slates the numbers you have mentioned; subtract 5 from each, and learn it by heart.

This done, questions are given by the teacher, and afterwards by the pupils, for practice. Next,

T. Is this all you have learnt by committing these answers to memory, or are there yet other numbers which you are now able to subtract as readily as 5?

P. We can likewise subtract 10+5, 20+5, 30+5, 40+5, 50, 51, 52, 53, 54, and 55.

T. Explain how it is, that by knowing how to subtract 5, you are able to subtract 10+5?

P. We have learnt to subtract 10 in the first lesson, and now we have learnt to subtract 5, therefore we must be able to subtract 10+5.

T. Explain how it is you are able to subtract 25, 35, 45, &c.

P. [Give similar reasons as above.]

Questions are now given by the teacher, and afterwards by the pupils, the results required to be explained, and finally, the exercises, Part II, are taken up for further practice.

Answers to the Exercises. Lesson V.

Ans. 1. 78, 73, 68, 63, 58, 53, 48, 43, 38, 33, &c.

2. 109, 94, 79, 64, 49, 34, 19, 4.

3. 206, 181, 156, 131, 106, 81, 56, 31, 6.

4. 337, 302, 267, 232, 197, 162, 127, 92, 57, 22.

5. 441, 396, 351, 306, 261, 216, 171, 126, 81, 36.

6. 534, 484, 434, 384, 334, 284, 234, 184, 134, 84.

7. 546, 495, 444, 393, 342, 291, 240, 189, 138, 87, 36.

8. 596, 544, 492, 440, 388, 336, 284, 232, 180, 128, 76, 24.

9. 844, 791, 738, 685, 632, 579, 526, 473, 420, 367, 314, 261, 208, 155, 102, 49.

10. 939, 885, 831, 777, 723, 669, 615, 561, 507, 453, 399, 345, 291, 237, 183, 129, 75, 21.

Ans. 11. 946, 891, 836, 781, 726, 671, 616, 561, 506, 451, 396, 341, 286, 231, 176, 121, 66, 11.

12. By adding the numbers.

Model.

13. 103+15=118, sum;
103−15=103−10−5=93−5=88, difference,

14. 233, sum; 183, difference.

15.	416	...	346	...
16.	462	...	372	...
17.	633	...	533	...
18.	701	...	599	...
19.	763	...	659	...
20.	879	...	769	...
21.	989	...	879	...
22.	1178	...	1068	...

Lesson VI.

To subtract 6, 16, 26, 36, 46, 56, 60, 61, 62, 63, 64, 65, and 66.

The pupils have to learn in this lesson to subtract 6 from 6, 7, 8......15; the rest is obvious.

Answers to the Exercises. Lesson VI.

Ans. 1. 94, 88, 82, 76, 70, 64, 58, 52, 46, 40, &c.

Ans. 2. 122, 106, 90, 74, 58, 42, 26, 10.
3. 153, 127, 101, 75, 49, 23.
4. 177, 141, 105, 69, 33.
5. 239, 193, 147, 101, 55, 9.
6. 263, 207, 151, 95, 39.
7. 334, 274, 214, 154, 94, 34.
8. 426, 365, 304, 243, 182, 121, 60.
9. 511, 449, 387, 325, 263, 201, 139, 77, 15.
10. 578, 515, 452, 389, 326, 263, 200, 137, 74, 11.
11. 718, 654, 590, 526, 462, 398, 334, 270, 206, 142, 78, 14.
12. 768, 703, 638, 573, 508, 443, 378, 313, 248, 183, 118, 53.
13. 851, 785, 719, 653, 587, 521, 455, 389, 323, 257, 191, 125, 59.

Ans. 14. 523 *Ans.* 16. 787 *Ans.* 18. 946
15. 647 17. 878 19. 1042

Lesson VII.

To subtract 7, 17, 27, 37, 47, 57, 67, 70, 71, 72, 73, 74, 75, 76, and 77.

Answers to the Exercises.

Ans. 1. 97, 90, 83, 76, 69, 62, 55, 48, 41, 34, 27, &c.
2. 142, 125, 108, 91, 74, 57, 40, 23.

Ans. 3. 166, 139, 112, 85, 58, 31, 4.
4. 177, 140, 103, 66, 29.
5. 225, 178, 131, 84, 37.
6. 284, 227, 170, 113, 56.
7. 325, 258, 191, 124, 57.
8. 373, 303, 233, 163, 93, 23.
9. 422, 351, 280, 209, 138, 67.
10. 502, 430, 358, 286, 214, 142, 70.
11. 588, 515, 442, 369, 296, 223, 150, 77, 4.
12. 674, 600, 526, 452, 378, 304, 230, 156. 82, 8.
13. 751, 676, 601, 526, 451, 376, 301, 226, 151, 76, 1.
14. 857, 781, 705, 629, 553, 477, 401, 325, 249, 173, 97, 21.
15. 923, 846, 769, 692, 615, 538, 461, 384, 307, 230, 153, 76.
16. Sum = 391; difference = 237.
17. Sum = 530; difference = 376.

Lesson VIII.

To subtract 8, 18, 28, 38, 48, 58, 68, 78, 80, 81, 82, 83, 84, 85, 86, 87, and 88.

Answers to the Exercises.

Ans. 1. 113, 105, 97, 89, 81, 73, 65, 57, 49, 41, 33, 25, 17, 9, 1.

Ans. 2. 175, 157, 139, 121, 103, 85, 67, 49, 31, 13.

3. 185, 157, 129, 101, 73, 45, 17.
4. 233, 195, 157, 119, 81, 43, 5.
5. 248, 200, 152, 104, 56, 8.
6. 275, 217, 159, 101, 43.
7. 319. 251. 183. 115, 47.
8. 366, 288, 210, 132, 54.
9. 470, 390, 310, 230, 150, 70.
10. 586, 505, 424, 343, 262, 181, 100, 19.
11. 691, 609, 527, 445, 363, 281, 199, 117, 35.
12. 754, 671, 588, 505, 422, 339, 256, 173, 90, 7.
13. 804, 720, 636, 552, 468, 384, 300, 216, 132, 48.
14. 858, 773, 688, 603, 518, 433, 348, 263, 178, 93, 8.
15. 906, 820, 734, 648, 562, 476, 390, 304, 218, 132, 46.
16. 913, 826, 739, 652, 565, 478, 391, 304, 217, 130, 43.
17. 1023, 935, 847, 759, 671, 583, 495, 407, 319, 231, 143, 55.

Lesson IX.

To subtract 9, 19, 29, 39, 49, 59, 69, 79, 89, 90, 91, 92, 93, 94, 95, 96, 97, 98, and 99.

Answers to the Exercises.

Ans. 1. 92, 83, 74, 65, 56, 47, 38, 29, 20, 11, 2.

2. 163, 144, 125, 106, 87, 68, 49, 30, 11.

3. 175, 146, 117, 88, 59, 30, 1.

4. 231, 192, 153, 114, 75, 36.

5. 307, 258, 209, 160, 111, 62, 13.

6. 324, 265, 206, 147, 88, 29.

7. 378, 309, 240, 171, 102, 33.

8. 474, 395, 316, 237, 158, 79, 0.

9. 576, 487, 398, 309, 220, 131, 42.

10. 641, 551, 461, 371, 281, 191, 101, 11.

11. 828, 737, 646, 555, 464, 373, 282, 191, 100, 9.

12. 857, 765, 673, 581, 489, 397, 305, 213, 121, 29.

13. 852, 759, 666, 573, 480, 387, 294, 201, 108, 15.

14. 906, 812, 718, 624, 530, 436, 342, 248, 154, 60.

15. 1007, 912, 817, 722, 627, 532, 437, 342, 247, 152, 57.

16. 1088, 992, 896, 800, 704, 608, 512, 416, 320, 224, 128, 32.

17. 1116, 1019, 922, 825, 728, 631, 534, 437, 340, 243, 146, 49.

18. 1143, 1045, 947, 849, 751, 653, 555, 457, 359, 261, 163, 65.

19. 1238, 1139, 1040, 941, 842, 743, 644, 545, 446, 347, 248, 149, 50.

N.B. In the subtraction of 9, 19, 29, &c., the pupils must be led to observe that it is easier to subtract 10, 20, 30, &c., and then to add 1 to the result.

Ans. 20.	Sum	=	107;	difference	=	31.
21.	...	=	143;	...	=	29.
22.	...	=	134;	...	=	44.
23.	...	=	208;	...	=	22.
24.	...	=	335;	...	=	139.
25.	...	=	428;	...	=	254.
26.	...	=	547;	...	=	349.
27.	...	=	678;	...	=	484.
28.	...	=	711;	...	=	515.
29.	...	=	866;	...	=	688.
30.	...	=	897;	...	=	739.
31.	...	=	1001;	...	=	867.
32.	...	=	1005;	...	=	755.
33.	...	=	914;	...	=	628.
34.	...	=	862;	...	=	464.
35.	...	=	759;	...	=	347.
36.	...	=	658;	...	=	224.
37.	...	=	677;	...	=	89.
38.	...	=	600;	...	=	28.
39.	...	=	593;	...	=	313.
40.	...	=	653;	...	=	185.
41.	...	=	1655;	...	=	179.

CHAPTER III.—MULTIPLICATION.

§ 1. MULTIPLICATION BY UNITS.

LESSON I. *To multiply by* 1 *and by* 2.

THE subject of multiplication might be treated like that of addition; and accordingly be begun by taking or repeating a number *once,* 1 ten times, 10+1 times, 1 hundred times, &c. The next step would be to repeat a number *twice,* 2 ten times, 20+1 times, 20+2 times, 2 hundred times, &c.; and so on with the numbers 3, 4, 5, &c. For, developed as the pupils' minds must be by the foregoing, little effort is required by them to conceive a number repeated 10 times; since to repeat 1 10 times is evidently making 1 unit, 1 ten; again, to repeat 2 10 times is to make 2, 2 tens, and to repeat or multiply 37 10 times is to make 37, 37 tens. This once understood, the multiplication by 11 or 10+1 will immediately result from it. Thus, to multiply a number by 11 is to repeat it 10 times, (which is known), and once more, (which is also known); again, if it be known how to multiply the numbers 1, 2, 3, ... 10, by 2, it follows, that the multiplication of 10+1, 10+2, 10+3, &c., by 2, is likewise known; and this obtained, the multiplication of any number by 10+2, *i. e.* by 12, is immediately deduced. Nothing prevents them from applying the same to the multiplication by 20 or 2 tens, for it is known how to multiply by 2; hence to multiply a number by 2 tens is to take it twice; but then the results are not units but tens. To multiply by 20+1 and 10+2 are the immediate results from this consideration. If, then, the teacher find his class sufficiently developed, it will much more tend to their further development if they pursue this course, than if they simply begin with multiplying a number by 2, next by 3, then by 4, &c. This healthy state of mind is, however, seldom found in all the pupils in a class, and then it is advisable to choose the simpler and slower process. The same objection ought not to deter the teacher from treating subtraction in the manner shown before; the results there being smaller numbers; whereas in multiplication their increase is sudden and great.

Teacher. How much is one 1?
How much is one 2?
How much is one 3?
How much is one 4?
How much is one 5?
&c., &c.
How much is one 87?

Pupils. 1, 2, 3, 4, 5, &c., 87.

T. How much is 2 ones?
How much is 2 twos?
How much is 2 threes?
How much is 2 fours?

P. 2, 4, 6, 8.

T. Instead of saying, "how much is 2 ones," we might say, "how much is 1 taken twice," or "how much is twice 1." How much is twice 5? twice 6? twice 7?

P. 10, 12, 14.

T. How much is twice 8? twice 9? twice 10?

P. 16, 18, 20.

T. Repeat together twice 1 is 2.
... ... twice 2 is 4.
... ... twice 3 is 6.
&c.
... ... twice 10 is 20.

P. [Repeat.]

T. Do all of you know this readily?—if so, can you tell how much twice 11 is?

P. 11 is 10 more 1; hence twice 11 is twice 10 more twice 1, which is 20 more 2, or 22.

T. How much is twice 12?

P. Twice 10 more twice 2, or 20 more 4, which is 24.

T. How much is twice 13?

P. Twice 10 more twice 3, or 20 more 6, which is 26.

T. How much is twice 14?
... ... twice 15?

T. How much is twice 16?
... ... twice 17?
... ... twice 18?
... ... twice 19?

P. Twice ten more twice 9, or 20 more 18, which is 38?

T. How much is twice 2?

P. 4.

T. How much is twice 2 tens or twice 20?

P. 4 tens or 40.

T. How much is twice 3?

P. 6.

T. How much is twice 3 tens or twice 30?

P. 6 tens or sixty.

T. How much is twice 4?

P. 8.

T. How much is twice 4 tens or twice 40?

P. 8 tens or 80.

T. How much is twice 5 tens or twice 50?
... ... twice 6 tens or twice 60?
... ... twice 7 tens or twice 70?
... ... twice 8 tens or twice 80?
.. ... twice 9 tens or twice 90?
... ... twice 10 tens or twice 100?

Class repeat: twice 20 is 40.
... twice 30 is 60.
... twice 40 is 80.
&c.
... twice 100 is 200.

T. Decompose 21 into tens and units.

P. 21 is 2 tens more 1.

T. How much is twice 21?

P. Twice 20 more twice 1, or 40 more 2, which is 42.

T. How much is twice 22?

P. Twice 20 more twice 2, or 40 more 4, or 44.

T. How much is twice 23?

... twice 24?

... twice 25, &c.

... twice 31? twice 32? &c.

... twice 47? twice 89?

P. Twice 80 more twice 9, or 160 more 18, which is 178.

T. How much is twice 147?

P. Twice 100, more twice 40, more twice 7, or 200 more 80, more 14, which is 294.

T. You have now learnt to take any number twice, or as it is usually called to *multiply by* 2. If you know how to take 1 twice, or to multiply 1 by 2, which other numbers do you then know how to multiply by 2?

P. 10, 11, 100, 101, 111, 1000, &c.

T. Why? explain.

P. Because twice 1 is 2; therefore twice 1 ten is 2 tens or 20.

And because 11 is 10 more 1, twice 11 is twice 10 more twice 1, or 22. Again, because twice 1 is 2, therefore twice 100 is 200, and twice 101 is twice 100 more twice 1, or 202;

and twice 111 is twice 100, more twice 10, more twice 1, or 222.

T. If you know how to multiply 2 by 2, what other numbers do you then know how to multiply by 2?

P. 20, 21, 22, 200, 201, 202, 221, 222, &c.

T. Explain.

P. [Give similar reasons as before.]

And so on with the other numbers.

T. In order then to multiply any number whatever by 2, it is sufficient to learn to multiply actually which numbers only?

P. The numbers 1, 2, 3, 4, 5, &c., 9.

T. What is the result called when two or more numbers are added?

P. Their sum.

T. In the written exercises what sign did we use to indicate addition?

P. The sign +.

T. What is the result called when a lesser number is subtracted from a greater: and what sign did we use to indicate subtraction?

P. Their difference. We used the sign –.

T. You have learnt to multiply numbers by 2; the result obtained by multiplication is called *product;* and the sign used to indicate multiplication is ×. What is the sum, difference, and product of the numbers 2 and 78?

P. Their sum is 80; their difference 76; and their product 156.

T. Write this on your slates, using the proper signs for addition, subtraction, and multiplication.

P. [Write] $78+2=80$ sum.

$78-2=76$ difference.

$78\times2=156$ product.

The pupils are now called upon to give questions to the class on the previous lesson, after which the exercises, Part II, are taken up for further practice.

Answers to the Exercises. Lesson I.

Models.

$2\times17=2\times10+2\times7=20+14=34.$

$2\times89=2\times80+2\times9=160+18=178.$

$2\times178=2\times100+2\times70+2\times8=200+140.$
$+16=356.$

Ans.	1.	34	*Ans.*	7.	158	*Ans.*	13.	414
	2.	56		8.	176		14.	556
	3.	72		9.	192		15.	698
	4.	98		10.	246		16.	974
	5.	114		11.	274		17.	1190
	6.	130		12.	378		18.	1978

Lesson II. *To multiply by 3.*

Teacher. How much is 2×1?

Pupils. 2.

T. How much more is 3×1.

P. 1 more.

T. How much then is 3×1?

P 3; $2+1$ or 3.

T. How much is 2×2?

P. 4.

T. How much more is 3×2?

P. 2 more.

T. How much then is 3×2?

P. $4+2$ or 6.

T. How much is 2×3?

P. 6.

T. How much more is 3×3?

P. 3 more.

T. How much then is 3×3?

P. $6+3$ or 9.

T. You have learnt to take any number twice or to multiply any number by 2, can you tell what must be done to take any number 3 times, or to multiply it by 3?

P. First multiply the number by 2, and then add the number to the product.

T. Hence you are able to multiply any number by 3; multiply 56 by 3.

P. $2\times56=112$; therefore $3\times56=112+56=168$.

T. Would it not be more convenient to be able to tell at once the product of any number by 3 without first multiplying it by 2? The products of which numbers is it sufficient to know in order to do this? [should this question not be answered in a satisfactory manner]—

T. [continues.] The products of which numbers is it sufficient to know in order to multiply any number by 2?

P. Of the numbers 1, 2, 3, &c., 9.

T. See whether it be sufficient to know the products of these same numbers by 3.

P. Yes; no ——

T. Suppose you know how much 3×1 is; can you tell how much 3×10 is?

P. Yes, for $3 \times 1 = 3$, therefore 3×1 ten $= 3$ tens, or $3 \times 10 = 30$.

T. And if you know how much 3×10 is, can you tell how much 3×11 is?

P. Yes, for $11 = 10 + 1$, therefore $3 \times 11 = 3 \times 10 + 3 \times 1 = 30 + 3 = 33$.

T. You know how much 3×2 is, and I think you are able to ascertain from this readily how much 3×20 is?

P. Yes; for $3 \times 2 = 6$, therefore 3×2 tens $= 6$ tens, or $3 \times 20 = 60$.

T. And now I think, too, you are able to tell how much 3×12 is.

P. Yes; for $12 = 10 + 2$, therefore $3 \times 12 = 3 \times 10 + 3 \times 2 = 30 + 6 = 36$.

T. There are yet at least two other numbers which you are able to multiply by 3, which are they?

P. 21 and 22; for $3 \times 21 = 3 \times 20 + 3 \times 1 = 60 + 3 = 63$; and $3 \times 22 = 3 \times 20 + 3 \times 2 = 60 + 6 = 66$.

T. What do you now think of my former question?

P. If we know the products of the numbers 1, 2, 3, ... 10, by 3, we shall be able to find readily those of any other number.

T. Then find the products of these 10 numbers; write them on your slates and learn by heart. When known—

Class repeat: $3\times 1=3.$
$3\times 2=6.$
$3\times 3=9.$
&c.
$3\times 10=30.$

T. Apply this to tens.

Class repeat: $3\times 10=30.$
$3\times 20=60.$
$3\times 30=90.$
&c.
$3\times 100=300.$

Here follow promiscuous questions given at first by the teacher, and then by the pupils; every answer, right or wrong, must be subjected to the decision of the class, a practice much recommended, as inducing the pupils to take a lively interest, and thereby sustaining their attention.

Example. How much is 3×89?

P. 267; for $3\times 89=3\times 80+3\times 9=240+27=267.$

Answers to the Exercises. Lesson II.

Ans.	1.	111	*Ans.* 7.	729	*Ans.*	13.	2064
	2.	177	8.	807		14.	2391
	3.	234	9.	1155		15.	2508
	4.	288	10.	1428		16.	2625
	5.	339	11.	1779		17.	2961
	6.	387	12.	1872		18.	3702

Lesson III. *To multiply by* 4.

According to the plan detailed in the previous lessons, the pupils must be led to ascertain,

1. That to multiply a number by 4, is to multiply it first by 3, and then to add the number to the product.

2. That it is sufficient to know the products of the units by 4, in order to multiply any number by 4. Hence the necessity of actually learning these products by heart.

3. This done, the results are immediately applied to the multiplication of tens by 4.

4. Questions are given by the teacher and each pupil, in turn.

5. The exercises, Part II, are taken up.

Here follow a few questions, with their solutions, such as the pupils must be able to give, before they begin the next lesson.

Question. How much is 4×39?

Answer. 156; for $4 \times 39 = 4 \times 30 + 4 \times 9$
$= 120 + 36 = 156$.

Q. What is the product of 87 and 4?

A. 348; for $4 \times 87 = 4 \times 80 + 4 \times 7$
$= 320 + 28 = 348$.

Q. Multiply 178 by 4.

A. 712; for $4 \times 178 = 4 \times 100 + 4 \times 70 + 4 \times 8$
$= 400 + 280 + 32 = 680 + 32 = 712$.

Answers to the Exercises. Lesson III.

Ans. 1.	188	*Ans.* 7.	556	*Ans.* 13.	1468
2.	236	8.	628	14.	1556
3.	296	9.	876	15.	1772
4.	348	10.	992	16.	2348
5.	392	11.	1108	17.	2712
6.	420	12.	1300	18.	3156

Lesson IV. *To multiply by 5.*

Answers to the Exercises.

Ans. 1.	430	*Ans.* 8.	1065	*Ans.* 15.	3390
2.	485	9.	1105	16.	3435
3.	540	10.	1745	17.	3945
4.	685	11.	2290	18.	3990
5.	795	12.	2425	19.	4495
6.	875	13.	2835	20.	4990
7.	920	14.	2880	21.	6170

LESSON V. *To multiply by* 6.

Answers to the Exercises.

Ans. 1.	462	*Ans.* 8.	1302	*Ans.* 15.	3534
2.	534	9.	1404	16.	4068
3.	690	10.	2070	17.	4134
4.	828	11.	2202	18.	4482
5.	894	12.	2736	19.	5256
6.	1122	13.	2868	20.	5922
7.	1194	14.	3402		

LESSON VI. *To multiply by* 7.

Answers to the Exercises.

Ans. 1.	455	*Ans.* 8.	1512	*Ans.* 15.	4032
2.	546	9.	1638	16.	4368
3.	623	10.	2415	17.	4746
4.	679	11.	2576	18.	5523
5.	728	12.	3381	19.	6279
6.	973	13.	3472	20.	6909
7.	1281	14.	3724	21.	6951

LESSON VII. *To multiply by* 8.

Answers to the Exercises.

Ans. 1.	344	*Ans.* 4.	672	*Ans.* 7.	984
2.	464	5.	768	8.	1512
3.	632	6.	912	9.	1704

Ans.	10. 1792	*Ans.*	15. 3032	*Ans.*	20. 4448
	11. 1880		16. 3288		21. 5336
	12. 1968		17. 3376		22. 6224
	13. 2856		18. 3472		23. 7112
	14. 2944		19. 4360		24. 7944

Lesson VIII. *To multiply by* 9.

Answers to the Exercises.

Ans.	1. 333	*Ans.*	9. 1044	*Ans.*	17. 3942
	2. 432		10. 1143		18. 4401
	3. 531		11. 2142		19. 4635
	4. 549		12. 2241		20. 4887
	5. 648		13. 2259		21. 6012
	6. 747		14. 3258		22. 6975
	7. 846		15. 3357		23. 8073
	8. 945		16. 3843		24. 8874

Lesson IX. *Continued Products.*

Teacher. How much is 2 × 2?

Pupils. 4.

T. And how much is 2 × 4?

P. 8.

T. How much, then, is 2 × 2 × 2?

P. 8.

T. How much is 2 × 3?

P. 6.

T. How much is 2×6?

P. 12.

T. How much, then, is $2 \times 3 \times 2$?

P. 12.

T. Can you tell what you have done here?

P. We have first multiplied 3 by 2, and again multiplied the product by 2.

T. Hence you have connected the numbers 2, 3, 2, by multiplying the first two, and then continued by multiplying the product by the next number. The answer in such a case is called the *continued product* of the three numbers. What is the continued product of 3, 4, 5?

P. 60; because $3 \times 4 = 12$ and $5 \times 12 = 60$.

T. What is the continued product of 4, 5, 6?

P. 120; because $4 \times 5 = 20$, and $6 \times 20 = 120$.

T. If a number be multiplied by 2, and the product be multiplied by 2, by what number, then, is the first number multiplied?

P. By 4; because it is first taken twice, and the product, then, is taken twice; hence the first number is taken 4 times, or it is multiplied by 4.

T. Instead, then, of multiplying 87 by 4, in the manner you have learnt, what might be done?

P. 87 might first be multiplied by 2, and the product again by 2.

T. Try it both ways.

P. $4 \times 87 = 4 \times 80 + 4 \times 7 = 320 + 28 = 348$.

Or, $4 \times 87 = 2 \times 87 \times 2$;

$2 \times 87 = 174$, and $2 \times 174 = 348$.

T. What might be done instead of multiplying a number by 6, in the manner you have learnt before?

P. First multiply by 2, and the product by 3.

The present subject can, at this stage, be only adverted to, as its thorough application depends on the pupils being able to decompose numbers into factors, as treated fully in a subsequent chapter. This lesson may be omitted, if the class be not properly developed.

§2. MULTIPLICATION BY TENS.

LESSON I. *To multiply by* 10, 20, and 30.

Teacher. How many tens are 10×1?

Pupils. 1 ten.

T. How many units is this?

P. 10 units.

T. How many tens are 10×2?

P. 2 tens, or 20 units.

T. How many tens are 10×3?

P. 3 tens, or 30 units.

T. How many tens are 10×4?

... ... 10×5?

... ... 10×6?

&c., &c.

... ... 10×11?

P. 4 tens, or 40 units.
5 tens, or 50 units.
6 tens, or 60 units.
11 tens, or 110 units.

T. Hence, what may be said in general of a number multiplied by 10?

P. It is as many tens as there are units in the number.

T. How much, then, is 10×56?

P. 56 tens, or 560.

T. How much is 2 tens $\times$ 1?

P. 2 tens, or 20.

T. How much is 2 tens $\times$ 2?

P. 4 tens, or 40.

T. How much is 20×3?

P. 6 tens, or 60.

T. How much is 20×4?
... ... 20×5?
... ... 20×6?
&c.
... ... 20×10?
... ... 20×11?

P. 8 tens, or 80.
10 tens, or 100.
12 tens, or 120.
20 tens, or 200.
22 tens, or 220.

T. If, then, a number is to be multiplied by 2 tens, or by 20, what may be said?

P. It is multiplying that number by 2, and calling the product *tens*, not units.

T. Hence, if you know how to multiply by 2, you are likewise able to multiply——

P. By 2 tens, or by 20.

T. Multiply 37 by 20.

P. $20 \times 37 = 2 \times 37$ tens $=$ 74 tens $=$ 740.

T. Multiply 89 by 20.

P. $20 \times 89 = 2 \times 89$ tens $=$ 178 tens $=$ 1780.

T. How many tens are 30×1?

P. 3 tens, or 30 units.

T. How many tens are 30×2?

P. 6 tens, or 60 units.

T. How many tens are 30×3?

... ... 30×4? &c.

... ... 30×12?

P. 9 tens, or 90 units.

12 tens, or 120 units.

36 tens, or 360 units.

T. Hence, what may be said of a number multiplied by 3 tens, or by 30?

P. It is multiplying that number by 3, and calling the product *tens*.

T. If, then, you know how to multiply by 3, you are able to multiply likewise——

P. By 3 tens, or by 30.

T. Multiply 45 by 30.

P. $30 \times 45 = 3 \times 45$ tens $=$ 135 tens $=$ 1350 units.

Answers to the Exercises. Lesson I.

Ans. 1.	930	*Ans.* 7.	640	*Ans.* 13.	450
2.	870	8.	1180	14.	990
3.	1100	9.	1480	15.	1710
4.	1530	10.	1760	16.	2520
5.	2170	11.	2740	17.	2970
6.	3710	12.	4960	18.	4110

Lesson II. *To multiply by* 40, 50, *and by* 60.

In this lesson, the pupils must be led to ascertain that,

1. Multiplying a number by 40, is to multiply it by 4, and to call the product tens.

2. Multiplying a number by 50, is to multiply it by 5, and to call the product tens.

3. Finally, multiplying a number by 60, is to multiply it by 6, and to call the product tens.

Answers to the Exercises.

1st, by 40.		2nd, by 50.		3rd, by 60.	
Ans. 1.	720	*Ans.* 1.	950	*Ans.* 1.	660
2.	1360	2.	1250	2.	1320
3.	2320	3.	2350	3.	2100
4.	2680	4.	3400	4.	3360
5.	2920	5.	4100	5.	3840
6.	3800	6.	5050	6.	5700

Lesson II. *To multiply by* 70, 80, *and* 90.

Here, again, the pupils have to ascertain that multiplying a number by 70, 80, and 90, is to multiply it by 7, 8, and 9 respectively, and to call the products tens.

Answers to the Exercises.

1st, by 70.			2nd, by 80.			3rd, by 90.		
Ans.	1.	910	*Ans.*	1.	1120	*Ans.*	1.	1350
	2.	1330		2.	2960		2.	1890
	3.	1890		3.	3280		3.	3150
	4.	2450		4.	3840		4.	4140
	5.	3430		5.	4320		5.	5130
	6.	3710		6.	5040		6.	6120
	7.	4690		7.	6160		7.	7110
	8.	5880		8.	7120		8.	7560
	9.	6440		9.	7840		9.	8910
	10.	6790		10.	7600		10.	9360

§ 3. MULTIPLICATION BY TENS AND UNITS.

Lesson I. *To multiply by* 11, 12......19.

In the foregoing lessons, the pupils have learnt to multiply by 1 and by 10; by 2 and by 10; by 3, 4, &c., 9, and by 10;—they are now to multiply by 11, 12, 13, &c.; that is, by 10+1, 10+2, 10+3, &c. The teacher then will ask: If a number is to be multiplied by 11, what must be done.

Pupils. The number must first be multiplied by 10, and then by 1, and the two products must be added.

T. Multiply 11 by ...

P. $11 \times 11 = 10 \times 11 + 1 \times 11 = 110 + 11 = 121$.

T. Multiply 17 by 11.

P. $11 \times 17 = 10 \times 17 + 1 \times 17 = 170 + 17 = 187$.

T. Multiply 89 by 11.

P. $11 \times 89 = 10 \times 89 + 1 \times 89 = 890 + 89 = 979$.

T. If a number is to be multiplied by 12, what must be done?

P. First multiply by 10, then by 2, and add the two products.

T. Multiply 12 by 12.

P. $12 \times 12 = 10 \times 12 + 2 \times 12 = 120 + 24 = 144$.

T. Multiply 13 by 12.

P. $12 \times 13 = 10 \times 13 + 2 \times 13 = 130 + 26 = 156$.

T. Multiply 98 by 12.

P. $12 \times 98 = 10 \times 98 + 2 \times 98 = 980 + 196 = 1080 + 96 = 1176$.

Similarly with the multiplication by 13, 14, 15, &c.

T. Multiply 19 by 19.

P. $19 \times 19 = 10 \times 19 + 9 \times 19 = 190 + 171 = 290 + 71 = 361$.

T. Multiply 67 by 19.

P. $19 \times 67 = 10 \times 67 + 9 \times 67 = 670 + 603 = 1273$.

T. If, instead of multiplying 67 by 19, in the manner you have done, I multiplied by 20; by how much would the product be more than is required?

P. By once 67.

T. What, then, may be done to obtain 19×67?

P. Multiply by 20, and from the product subtract 67; thus:—

$$19 \times 67 = 20 \times 67 - 1 \times 67.$$
$$= 1340 - 67 = 1273.$$

T. I leave it to you to choose either of these two methods.

Teacher and pupils give promiscuous questions, after which the exercises, Part II, are taken up.

Answers to the Exercises.

By 11.

Ans.	1.	539	*Ans.*	3.	1089	*Ans.*	5.	1573
	2.	957		4.	1331		6.	2387

		By 12.	By 13.	By 14.
Ans.	1.	444	377	266
	2.	696	611	490
	3.	828	819	644
	4.	888	1092	798
	5.	996	1235	952
	6.	1164	1287	1106

		By 15.	By 16.	By 17.	By 18.	By 19.
Ans.	1.	255	288	323	288	285
	2.	420	464	357	486	494
	3.	585	496	544	684	703
	4.	615	672	731	882	912
	5.	780	848	918	918	1121
	6.	945	1024	1105	1116	1159
	7.	1110	1200	1292	1314	1368
	8.	1275	1376	1479	1512	1577
	9.	1440	1552	1666	1710	1786

LESSON II. *To multiply by* 21, 22......29.

Teacher. You have learnt to multiply by 20, and also by 1; if, then, you have to multiply a number by 21, what must be done?

Pupils. First multiply by 20, then by 1, and add the two products.

T. Multiply 27 by 21.

P. $21 \times 27 = 20 \times 27 + 1 \times 27.$

$= 540 + 27 = 567.$

The pupils must give a similar account as to the multiplication by 22, 23......29.

Answers to the Exercises.

		By 21.	By 22.	By 23.	By 24.
Ans.	1.	525	594	644	600
	2.	777	836	1035	864
	3.	1008	1078	1334	1128
	4.	1176	1254	1472	1392
	5.	1407	1606	2001	1656
	6.	1869	2156	2208	1800

		By 25.	By 26.	By 27.	By 28.	By 29.
Ans.	1.	625	884	729	1232	841
	2.	950	1144	1107	1568	1102
	3.	1225	1430	1566	2072	1363
	4.	1325	1716	1863	2324	1682
	5.	1875	2002	2106	2632	2001
	6.	2175	2288	2538	2772	2581

LESSON III. *To multiply by* 31, 32......39.

Teacher. Multiply 31 by 31.

Pupils. $31 \times 31 = 30 \times 31 + 1 \times 31 = 930 + 31 = 961.$

T. Multiply 87 by 32.

P. $32 \times 87 = 30 \times 87 + 2 \times 87 = 2610 + 174 = 2784.$

Answers to the Exercises.

	By 31.	By 32.	By 33.	By 34.
Ans. 1.	961	2784	2541	2278
2.	1302	3136	2904	2652
3.	1333	1024	3267	3026
4.	1674	1408	1122	3094
5.	2015	1760	1485	1122
6.	2356	2112	1848	1564

	By 35.	By 36.	By 37.	By 38.	By 39.
Ans. 1.	1995	1656	1332	3572	1248
2.	2380	2052	1739	3230	1677
3.	2765	2448	2146	2888	2106
4.	2835	2844	2553	2470	2535
5.	3220	2952	2627	2052	2964
6.	1225	3312	3071	1634	3393

LESSON IV. *To multiply by* 41, 42......49.

	By 41.	By 42.	By 43.	By 44.
Ans. 1.	1681	1764	1849	1936
2.	2132	2226	2322	2420
3.	2583	2688	2795	2904
4.	3034	3150	3268	3388
5.	3485	3612	3741	3872
6.	3936	4074	4214	4356

	By 45.	By 46.	By 47.	By 48.	By 49.
Ans. 1.	1710	4370	4183	3648	3087
2.	2205	1656	4277	4176	3526
3.	2295	2162	1504	4704	4165
4.	2790	2668	2021	1872	4704
5.	3285	3082	2538	1968	4263
6.	3780	3588	3055	2496	3724

Lesson V. *To multiply by* 51, 52......59.

Answers to the Exercises.

	By 51.	By 52.	By 53.	By 54.
Ans. 1.	2652	3536	3975	4428
2.	3213	4108	4558	5022
3.	3774	4212	5141	2916
4.	4335	4784	3074	3510
5.	4896	2756	3657	4104
6.	2907	3328	3763	4698

	By 55.	By 56.	By 57.	By 58.	By 59.
Ans. 1.	5390	3080	3534	4524	3835
2.	3245	3696	4161	5162	4484
3.	3355	4312	4788	5684	5015
4.	3960	4928	5415	4466	5546
5.	4565	5544	3192	3770	5723
6.	5170	2856	3819	3132	5841

LESSON VI. *To multiply by* 61, 62,......69.

	By 61.	By 62.	By 63.	By 64.
Ans. 1.	3721	2418	5922	1408
2.	4392	2542	5355	1216
3.	5063	3224	4788	1792
4.	5734	3902	4095	1984
5.	1037	4588	3402	2688
6.	1708	5270	2079	3392

	By 65.	By 66.	By 67.	By 68.	By 69.
Ans. 1.	4160	4356	1541	6052	2967
2.	4875	3630	2278	6664	1518
3.	5590	2904	3015	5916	1311
4.	6305	2178	3752	5168	2415
5.	5720	1452	4489	4420	3174
6.	5005	726	5226	3672	3933

LESSON VII. *To multiply by* 71, 72,......79.

	By 71.	By 72.	By 73.	By 74.
Ans. 1.	4828	4680	2117	6364
2.	5609	4032	2263	7178
3.	5751	3384	3066	7326
4.	6532	2736	3869	6438
5.	5893	2088	4672	5624
6.	5254	1296	5475	4810

	By 75.	By 76.	By 77.	By 78.	By 79.
Ans. 1.	4050	2356	7469	2574	5293
2.	3375	2192	7546	1716	6162
3.	2700	4028	6699	1794	7031
4.	2025	4864	5852	2652	7821
5.	1350	5700	4235	3510	7663
6.	2175	6536	3388	4368	6715

LESSON VIII. *To multiply by* 81, 82,......89.

Answers to the Exercises.

	By 81.	By 82.	By 83.	By 84.
Ans. 1.	6561	6232	2739	7980
2.	7452	5330	3735	8148
3.	7695	4428	4731	7224
4.	7857	3526	5727	6300
5.	8019	2624	5893	5376
6.	7047	1722	6889	4452

	By 85.	By 86.	By 87.	By 88.	By 89.
Ans. 1.	4165	8084	7569	7744	7921
2.	4845	8342	5829	8712	8722
3.	5525	8170	4959	3872	3293
4.	6205	7224	8439	6776	5162
5.	6885	6622	7221	5808	5607
6.	7905	7568	6873	4840	6853

Lesson IX. *To multiply by* 91, 92,......99.

Answers to the Exercises.

	By 91.	By 92.	By 93.	By 94.
Ans. 1.	8281	7176	8649	8836
2.	8463	6164	9021	9118
3.	8645	4968	8184	8084
4.	8827	3588	7161	4042
5.	9009	4508	6417	3478
6.	7735	5428	4464	5452

	By 95.	By 96.	By 97.	By 98.	By 99.
Ans. 1.	9025	9216	9409	9604	9801
2.	9405	8352	8439	9702	8712
3.	8265	5664	6499	1862	7623
4.	6270	4128	4656	3724	6534
5.	6935	3552	5626	5194	5445
6.	6555	2784	6596	6566	4356

Lesson X. *Promiscuous Exercises.*

Answers to the Questions.

Ans. 1.	495	*Ans.* 4.	224	*Ans.* 7.	5600
2.	324	5.	989	8.	3822
3.	376	6.	1312	9.	666

CHAPTER IV.—DIVISION.

To divide one number by another, is to see how often the latter is contained in the former. To commence with the simplest questions, the pupils must first ascertain how often 1, 2, 3, 4, &c., are contained in any number whatever. It is obvious, however, that if it be known how often 1, 2, 3, 4, &c., are contained in all numbers from 1 to 10, 1 to 20, 1 to 30, 1 to 40, &c., respectively, it may thence easily be deduced how often the same are contained in numbers exceeding these. To lead the pupils to the clear view of these truths, and to facility in calculating questions in reference to them, is the object of this chapter.

Lesson I. *To divide by* 1 *and by* 2.

Teacher. How many ones are there in
1, 2, 3, 4, ... 10?

Pupils. 1, 2, 3, 4, ... 10 ones.

T. Instead of saying how many ones are there in 1, 2, 3, &c., I might have asked you, how often 1 is contained in 1, in 2, in 3, &c. How often is 1 contained in 4, 5, 6, &c.?

P. 4 times, 5 times, 6 times, &c.

T. How will you convince me that 1 is contained in 2 twice?

P. $2 = 1 + 1$, and 1 is contained in 1, once, and in another 1, once more, that is, in 2, 1 is contained twice.

T. Try to express in words what you have done.

P. We distributed 2 into ones, and counted how often it was possible to do so.

T. Now reflect one moment; how many times must 1 be contained in any number whatever?

P. *As often as there are ones or units in that number.*

T. If now I ask you to see how often 2 is contained in 18, what will you do?

P. Distribute 18 into as many twos as possible, and count them thus:—

$$18 = 2 + 2 + 2 + 2 + 2 + 2 + 2 + 2 + 2,$$

That is, 2 is contained in 18, 9 times.

T. Instead of distributing 18 into twos, we might say *divide* 18 into twos; hence seeing how often 2 is contained in a number, *and dividing* it by 2, are the same thing. What does "to divide 48 by 2" mean?

P. To see how often 2 is contained in 48.

T. What does "to divide a number by 3, by 4, by 10" mean?

P. To see how often 3, 4, 10, are contained in that number.

T. We will now learn to divide any number by 2. In which number is 2 contained once?

P. In 2.

T. Twice?

P. In 2 twos, that is in 4.

T. In what number is 2 contained 3 times, 4 times, 5 times, 6 times, 7 times, 8 times, 9 times, 10 times?

P. In 3 twos or in 6;

4	...	8;
5	...	10;
6	...	12;
7	...	14;
8	...	16;
9	...	18;
10	...	20;

T. How much then is 2 divided by 2?

P. 1.

T. 4 divided by 2?

6	...	2?
8	...	2?
10	...	2?
12	...	2?
14	...	2?
16	...	2?
18	...	2?
20	...	2?

P. 2, 3, 4, 5, 6, 7, 8, 9, 10.

T. The sign used for division is ÷, which signifies "divided by;" thus [writing on the slate]

$$18 \div 2 = 9.$$

In this instance what does the number 9 indicate?

P. How often 2 is contained in 18.

T. The result obtained by adding numbers is called ——?

P. Their sum.

T. The result obtained by subtracting one number from another ——?

P. Their difference.

T. The result obtained by multiplication ——?

P. Their product.

T. And the result obtained by dividing one number by another is called a *quotient.*

In the example $18 \div 2 = 9$, which is the quotient?

P. 9.

T. The number by which we divide is called the *divisor,* and the other which we divide, the *dividend;* which is the divisor, and which is the dividend?

P. 2 is the divisor, and 18 the dividend.

T. Repeat the numbers which we have divided by 2.

P. 2, 4, 6, 8, 10, 12, 14, 16, 18, 20.

T. Which are the intermediate numbers to those?

P. 3, 5, 7, 9, 11, 13, 15, 17, and 19.

T. Divide these by 2, what are the quotients?

P. 3÷2=1 and 1 over.

T. Yes, 2 is contained in 3 once, and there remains 1; the number which remains after division is thence called the *remainder:* proceed.

P. 5÷2=2, with the remainder 1.
7÷2=3, ... 1.
&c., &c.

T. Now compare the quotients obtained by dividing 2, 4, 6, 8, 10, &c., by 2, and those obtained by dividing

3, 5, 7, 9, 11, &c., by 2,

and tell me what you remark.

P. The 2, 4, 6, 8, 10, &c., divided by 2, leave no remainder; whereas 3, 5, 7, &c., divided by 2, leave a remainder 1.

T. Hence we may say that some numbers are exactly divisible by 2, and others are not;—the former are called *even*—the latter *odd.* Now write on your slates all numbers, from 1 to 20; divide each by 2, and learn the quotients by heart.

P. [Write.] 2÷2=1.
3÷2=1, remainder 1.
4÷2=2.
5÷2=2, rem. 1.
&c., &c.

When known,

T. [Proceeds.] How much is

4 divided by 2 ? *Ans.* 2.
4 tens, or 40 ... 2 ? *Ans.* 2 tens, or 20.
40 tens, or 400 ... 2 ? *Ans.* 20 tens, or 200.
400 tens, or 4000 ... 2 ? *Ans.* 200 tens, or 2000.

6 divided by 2 ? *Ans.* 3.
6 tens, or 60 ... 2 ? *Ans.* 3 tens, or 30.
60 tens, or 600 ... 2 ? *Ans.* 30 tens, or 300.
600 tens, or 6000 ... 2 ? *Ans.* 300 tens, or 3000.

Again, How much is

8 divided by 2 ? *Ans.* 4.
8 tens, or 80 ... 8 ? *Ans.* 4 tens, or 40.
80 tens, or 800 ... 2 ? *Ans.* 40 tens, or 400.
800 tens, or 8000 ... 2 ? *Ans.* 400 tens, or 4000.

How much is

10 divided by 2 ? *Ans.* 5.
10 tens, or 100 ... 2 ? *Ans.* 5 tens, or 50.

How much is

12 divided by 2 ? *Ans.* 6.
12 tens, or 120 ... 2 ? *Ans.* 6 tens, or 60.

How much is

14 divided by 2 ? *Ans.* 7.
14 tens, or 140 ... 2 ? *Ans.* 7 tens, or 70.

How much is

16 divided by 2 ? *Ans.* 8.
16 tens, or 160 ... 2 ? *Ans.* 8 tens, or 80.

How much is

18 divided by 2? *Ans.* 9.

18 tens, or 180 ... 2? *Ans.* 9 tens, or 90.

How much is

20 divided by 2? *Ans.* 10.

20 tens, or 200 ... 2? *Ans.* 10 tens, or 100.

What remark have you to make?

P. That if it be known how to divide

2, 4, 6, 8, 10, 12, 14, 16, 18, 20,

by 2, it is easy to ascertain the quotients of

20, 40, 60, 80, 120, 140, 160, 180, 200;

of 200, 400, 600, &c., by 2.

T. Which are the intermediate tens to these numbers?

P. 30, 50, 70, 90, &c.

T. How much is 20 divided by 2?

P. 10.

T. How much is 10 divided by 2?

P. 5.

T. Then how much is 20+10, or 30, divided by 2?

P. 10+5, or 15.

T. How much is 50 divided by 2?

P. $50 = 40 + 10$; $40 \div 2 = 20$; $10 \div 2 = 5$. Hence 50 or $40+10 \div 2 = 20+5 = 25$.

T. How much is 70, 90, 110, 130, &c., divided by 2?

P. 35, 45, 55, 65, &c.

T. How much is $21 \div 2$?

P. 10, with remainder 1.

T. How much is 22÷2?

P. 22 = 20+2; 20÷2 = 10; and 2÷2 = 1. Hence 22÷2=10+1=11.

T. How much is 23, 24, 25, &c.,÷2?

P. 11, *r*. 1; 12; 12, *r*. 1; &c.

T. How much is 31, 32, 33, 34, &c.,÷2?

P. 34÷2 is 30÷2, and 4÷2;
30÷2=15, and 4÷2=2.
Hence 34÷2=17.

T. [Proceeds in a similar manner with the numbers 41, 42, 43, &c.; 51, 52, &c.] How much is 97÷2?

P. 97÷2 is 90÷2, and 7÷2;
90÷2=45, and 7÷2=3, rem. 1.
Hence 97÷2=45+3, rem. 1 = 48, rem. 1.

T. How much is 159÷2?

P. 159=100+50+9, which, divided by 2,
= 50+25+4, rem. 1.
Hence 159÷2=79, rem. 1.

T. If, then, you know how to divide certain numbers by 2, it is easy to ascertain the quotients of all other numbers by 2. Which are those certain numbers?

P. 2, 3, 4, 5, 6, 7, &c., 20.

The pupils are now called upon to give questions to the class, after which the exercises in Part II. are taken up.

Answers to the Exercises.

1. The number to be divided is called the *dividend;* the number which is contained in the dividend is called the *divisor;* and the number which shows how often the divisor is contained in the dividend is called the *quotient.*

2. The numbers themselves.

3. Even numbers are such as are exactly divisible by 2; *odd* numbers are such as are not exactly divisible by 2.

4. If odd numbers be divided by 2, the remainders are always 1.

5. *Ans.* 4.

6. That if a number be multiplied by 2, and the product be divided by 2, the quotient is always the number itself.

7. That if a number be divided by 2, and the quotient be multiplied by 2, the product is always the number itself.

Ans. 8. 94, *r.* 1; 118, *r.* 1; 175, *r.* 1; 200 *r.* 1; 253, *r.* 1; 300, *r.* 1; 391, *r.* 1; 448, *r.* 1; 494; 500, *r.* 1.

9. 186, *r.* 1.

10. 44.

11. 195, *r.* 1.

Ans. 12. 204.

13. 0.

LESSON II. *To divide by* 3, 4, *and* 5.

The division by 3, 4, 5, and succeeding numbers, is to be treated precisely in the same way as shown in the preceding lesson. The pupils, accordingly, are to find in which numbers 3 is contained once, twice, 3 times, 4 times......10 times (these are said to be divisible by 3.) Next, to find the quotients of all numbers between 3 and 30, and to *commit them to memory.* Then follows an important exercise:—

3 is contained in 3 .... once.
... ... 3 tens or in 30... 1 ten times.
... ... 30 tens or in 300... 10 ten times, or 100 times.

Again,
3 is contained in 6 .... twice.
... ... 6 tens, or 60 ... 2 tens, or 20 times.
... ... 60 tens, or 600 ...20 tens, or 200 times.

Similarly,
3 is contained in 9 .... 3 times.
... ... 90 .... 30 times.
... ... 900 ....300 times.
... ... 12 .... 4 times.
... ... 120 .... 40 times.
... ... 1200 ....400 times.
&c.

These exercises are important, as they give the pupils facility in dividing the intermediate tens, 40, 50, 70, 80, 100, &c.; for, since they know that 3 is contained in 30 10 times, the question, how often it is contained in 40, will present no new difficulty; they will naturally conclude that 3 is contained in 30+10, 10+3 times, with remainder 1; similarly 3 is contained in 50, that is in 30+20, 10+6 times, with the remainder 2; again, 3 is contained in 70, that is in 60+10, 20+3 times, with the remainder 1; and so on with other numbers, which, taught by the above exercises, the pupils will separate each time into two or more convenient numbers: divide each of them and add the quotients. For instance, the question, "Divide 87 by 3," will be solved thus, by pupils who have gone slowly and regularly through the above exercises: viz.

3 is contained in 60...... 20 times,
and in the remainder 27...... 9 times.

Hence, $87 \div 3 = 29$.

Again, Divide 158 by 3.

Solution. $150 \div 3 = 50$.

$8 \div 3 = 2$, *r.* 2.

Hence $158 \div 3 = 52$, *r.* 2.

And again, Divide 259 by 3.

Solution.

$240 \div 3 = 80$; i. e. 24 tens $\div$ 3 $=$ 8 tens;

and $19 \div 3 = 6$, *r.* 1.

Hence $259 \div 3 = 86$, *r.* 1.

Teacher and pupils give questions to the class, and the exercises in Part II conclude division by 3. The steps in dividing by 4, 5, 6, are precisely the same. The pupils have to *learn by heart* the quotients of the numbers 4 to 40, 5 to 50, 6 to 60, by 4, 5, 6, respectively;—these known, there follow what were called above, *important* exercises. Questions by the teacher and pupils, and the exercises in Part II, conclude each lesson.

Answers to the Exercises.

By 3.

Ans. 1. 3, 6, 9, 12, 15, 18, 21, 24, 27, 30, 33, 36, 39, &c. .... 99.

2. Either 1 or 2.

3. No, for if the remainder be 3, the quotient will be 1 more; the same with 4 or 5.

4. 38, *r.* 1; 84, *r.* 1; 127; 135, *r.* 2; 175, *r.* 2; 229, *r.* 1; 233, *r.* 2; 276, *r.* 1; 328; 335, *r.* 2; 780; 1033, *r.* 2.

5. 152. *Ans.* 6. 252.

7. 1260.

By 4.

Ans. 1. 4, 8, 12, 16, 20, 24, 28, 32, 36, &c. 100.

Ans. 2. Either 1, or 2, or 3.

3. There cannot be a remainder 4, or more than 4, because 4 would be contained in such remainders.

4. 34; 54, *r.* 3; 76, *r.* 1; 118, *r.* 1; 147, *r.* 1; 169, *r.* 1; 197, *r.* 1; 212, *r.* 3; 227, *r.* 1; 253; 328, *r.* 3; 505, *r.* 1.

5. 104, *r.* 3.

Ans. 6. 118, *r.* 2.

7. 1836.

By 5.

Ans. 1. 5, 10, 15, 20, 25, &c.

2. Either 1, 2, 3, or 4.

3. 29, *r.* 3; 54, *r.* 1; 60, *r.* 4; 96, *r.* 3; 119, *r.* 3; 123, *r.* 2; 148, *r.* 1; 171, *r.* 4; 189, *r.* 2; 202, *r.* 1; 429, *r.* 4; 655, *r.* 3; 823, *r.* 1; 1065, *r.* 4.

4. 107

5. 46, *r.* 2.

6. 80.

7. 46, *r.* 1.

Ans. 8. 88.

9. 87.

10. 261.

LESSON III. *To divide by* 6, 7, 8, 9.

1. The pupils have to *commit to memory* the quotients of the numbers, 6, 7, 8 ...... 60 by 6;
... ... 7, 8, 9 ...... 70 by 7;
... ... 8, 9, 10 ...... 80 by 8;
... ... 9, 10, 11 ...... 90 by 9.

2. Then follow *important* exercises, viz.

6 divided by	6	....	= 1.
60 ...	6	....	= 1 ten.
600 ...	6	....	= 1 hundred.
	&c.		

12 divided by	6	....	= 2.
120 ...	6	....	= 2 tens.
1200 ...	6	....	= 2 hundreds.
	&c.		

Likewise,

7 divided by	7	....	= 1.
70 ...	7	....	= 1 ten.
700 ...	7	....	= 1 hundred.
	&c.		

And 8 divided by	8	....	= 1.
80 ...	8	....	= 1 ten.
800 ...	8	....	= 1 hundred.
	&c.		

Also, 9 divided by	9	....	= 1.
90 ...	9	....	= 1 ten.
900 ...	9	....	= 1 hundred.
	&c.		

3. Questions by the teacher and pupils to the class.

4. The exercises in Part II, conclude each lesson.

Answers to the Exercises.

By 6.

Ans. 1. 6, 12, 18, 24, 36, 42, 48, 54, 60, 66, &c.

Ans. 2. Either 1, 2, 3, 4, or 5.

3. 23; 34, *r.* 1; 64, *r.* 5; 78, *r.* 5; 97, *r.* 4; 111, *r.* 1; 149, *r.* 5; 148; 156, *r.* 4; 167, *r.* 1; 352, *r.* 2; 535; 720, *r.* 1.

Ans. 4. 28. *Ans.* 6. 192, *r.* 5.

5. 28. 7. 43, *r.* 1.

By 7.

Ans. 1. 7, 14, 21, 28, 35, 42, 49, 56, 63, 70, &c.

2. Either 1, 2, 3, 4, 5, or 6.

3. 16, *r.* 2; 32, *r.* 1; 48; 65, *r.* 1; 81; 96, *r.* 6; 127; 141; 176, *r.* 2; 235; 493, *r.* 5.

Ans. 4. 29, *r.* 1. *Ans.* 6. $3 \times 4 \times 5 \times 6 = 360$.

5. 55, *r.* 4. 7. 1016.

By 8.

Ans. 1. 8, 16, 24, 32, 40, 48, 56, 64, &c.
2. Either 1, 2, 3, 4, 5, 6, or 7.
3. 19, *r.* 7; 32, *r.* 4; 46, *r.* 3; 60, *r.* 2; 74, *r.* 1; 85, *r.* 4; 95, *r.* 3; 109, *r.* 7; 123, *r.* 3; 140, *r.* 2; 293, *r.* 1.

Ans. 4. 33, *r.* 2.
5. 61, *r.* 1.

Ans. 6. $7 \times 9 \times 10 = 630$.
7. 1380.

By 9.

Ans. 1. 9, 18, 27, 36, 45, 54, 63, 72, 81, 90, &c.
2. Either 0, 1, 2, 3, 4, 5, 6, 7, or 8.
3. 15, *r.* 1; 27, *r.* 4; 39, *r.* 7; 52, *r.* 1; 63, *r.* 3; 75, *r.* 6; 88; 99, *r,* 2; 109, *r.* 8; 108, *r.* 6; 146.

Ans. 4. 8, *r.* 6.
5. 58, *r.* 6.

Ans. 6. $8 \times 10 \times 11 = 880$.
7. 1507.

Lesson IV. *To divide by* 10, 11, 12.

Answers to the Exercises.

By 10.

Ans. 1. 10, 20, 30, 40, &c.

Ans. 2. 73 is not divisible by 10, because 3 remains after division. 130 is divisible by 10, because 0 remains after division. All numbers ending in 0, are divisible by 10.

3. All numbers of which the unit figure is either 0, 2, 4, 6, or 8, are divisible by 2. All numbers of which the unit figure is either 0 or 5, are divisible by 5.

4. 31, *r.* 7; 58, *r.* 6; 61; 79, *r.* 3; 89, *r.* 7; 99, *r.* 9; 100; 101; 220; 307.

By 11.

Ans. 1. 11, 22, 33, 44, &c.

2. 9, *r.* 5; 19, *r.* 8; 29, *r.* 7; 50, *r.* 5; 57, *r.* 6; 67, *r.* 7; 77, *r.* 8; 90, *r.* 9; 91, *r.* 10.

3. 13, *r.* 8.

4. How much is the sum of 58 and 93, divided by 11?

5. 36.—Divide 66 by 11, and multiply the quotient by 6.

6. 40, *r.* 4. Divide the difference of 731 and 287 by 11.

7. 280.

By 12.

Ans. 1. 12, 24, 36, 48, 60, 72, 84, 96, 108, 120, 132, 144, 156, 168, 180, 192.

2. 12, *r.* 9; 23, *r.* 2; 29, *r.* 11; 37, *r.* 4; 47, *r.* 3; 64, *r.* 5; 74, *r.* 6; 76, *r.* 10; 83, *r.* 4.

3. 17, *r.* 9. Divide the sum of 59, 67, and 87, by 12.

4. 63, *r.* 10. Divide the difference between 1139 and 373, by 12.

5. 172, *r.* 6. Divide the product of 345 and 6, by 12.

6. 64.—To the quotient of 180 by 12, add 49.

7. $12 \times 12 = 144$.—Divide the continued product of 12, 12, and 12, by 12.

N.B. The method exhibited in these lessons applies to the division by the numbers, 13, 14, 15, &c. The limit to these exercises is that where the pupils cease to conceive with clearness, and calculate with facility; yet it is advisable to require the pupils to divide all numbers up to 100 or 200, by any number not exceeding these.

Another useful exercise is the following:—a number is given, which it is required to divide by 2, 3, 4, 5, &c., successively. The pupils may exhibit a similar question on their slates, thus:—

Quest. Divide 750 by 2, 3, 4, 5......9.

Dividend 750.	Divisors,	2.	3.	4.	5.
	Quotients,	375	250	187, *r.* 2	150

Dividend 750.	Divis.	6.	7.	8.	9.
	Quot.	125	107, *r.* 1	93, *r.* 6	83, *r.* 3

CHAPTER V.

MULTIPLES, AND DIVISORS OF NUMBERS—PRIME NUMBERS—SQUARE NUMBERS.

LESSON I. *Multiples and Divisors.*

THE subjects of this chapter are important, as they are a preparation for fractions. The operations to be performed are easy; the object is to gain clear ideas, and these obtained, to express them in proper terms.

Teacher. If a number contains another number exactly, what may be said of those numbers?

Pupils. The first number is *divisible* by the second.

T. Give an instance.

P. 20 is divisible by 10.

T. What operation must be performed upon the number 10 in order to obtain 20?

P. 10 must be *added* to it; or 10 must be *multiplied* by 2.

T. Give another instance of a number divisible by another.

P. 60 is divisible by 5.

T. What operation must, in this instance, be performed upon the number 5 in order to obtain 60?

P. 5 must be added 11 times to 5, or 5 must be multiplied by 12.

T. It is for this reason that 60 is said to be a *multiple* of 5. Tell me now of which other number, besides 5, 60 is a *multiple.*

P. 60 is a multiple of 1, of 2, or of 3, or of 4, of 6, of 10, of 12, of 15, of 30.

T. Why?

P. Because 60 is obtained by multiplying these numbers by 60, 30, 20, 15, 10, 6, 5, 4, and 2 respectively.

T. Name two numbers, of which the one may be said to be a *multiple* of the other.

P. 12 and 3, 16 and 4, &c.

T. Try to express in general the meaning of the word *multiple.*

P. A number which contains another number a certain number of times exactly, is said to be a *multiple* of the latter.

T. Name two numbers, of which the one is not a multiple of the other, and state the reason why it is not.

P. 18 is not a multiple of 7, because 18 does not contain 7 exactly.

T. Find 6 numbers which are multiples of 9, 7, 10, 13, and 15.

P. Multiples of 9 are 18, 27, 36, 45, &c.
... of 7 are 14, 21, 28, 35, &c.
... of 10 are 20, 30, 40, &c.
... of 13 are 26, 39, 52, &c.
... of 15 are 30, 45, 60, &c.

T. Find all numbers of which 100 is a multiple.

P. 100 is a multiple of 1, 2, 4, 5, 10, 20, 25, and 50.

T. In order to answer this last question, what were you obliged to do?

P. Find all numbers which are exactly contained in 100, or by which 100 is divisible.

T. It is for this reason these numbers are said to be *divisors* of 100. Name a number which is a divisor of 12.

P. 1, or 2, or 3, or 4, or 6, or 12.

T. Why?

P. Because each of these numbers is contained exactly in 12.

T. Name two numbers, of which the one is not a divisor of the other; and state the reason why it is not.

P. 5 is not a divisor of 12, because it is not contained in 12 exactly.

T. Find some numbers which are divisors of 15, 20, 40, 90, 3, and 7.

P. Divisors of 15, are 1, 3, 5, and 15.
... 20, ... 1, 2, 4, 5, 10, and 20.
... 40, ... 1,2, 4, 5, 8, 10, 20, and 40.

Divisors of 90, are 1, 2, 3, 5, 6, 9, 10, 18, 30, 45, and 90.

... 3, ... 1 and 3.

... 7, ... 1 and 7.

Answers to the Exercises.

Ans. 1. A number which contains another number a certain number of times exactly, is said to be a multiple of the latter.

Ans. 2. A number which does not contain another number a certain number of times exactly, is said to be not a multiple of the latter.

Ans. 3. All numbers are multiples of 1.

Ans. 4. Multiples of 13, are 26, 39, 52, &c.

... 27, ... 54, 81, 108, &c.

... 59, ... 118, 177, 236, &c.

... 87, ... 174, 261, 348, &c.

Ans. 5. 108.

Ans. 6. 104 is a multiple of 1, 2, 4, 8, 13, 26, 52, 104.

Ans. 7. A number which is contained in another number exactly, is said to be a divisor of the latter.

Ans. 8. A number which is not contained in another number exactly, is said not to be a divisor of the latter.

Ans. 9. 1 is a divisor of every number.

Ans. 10. Divisors of 28, are 1, 2, 4, 7, 14, 28.
... 39, ... 1, 3, 13, 39.
... 48, ... 1, 2, 3, 4, 6, 8, 12, 16, 24, 48.
... 17, ... 1, 17.
... 29, ... 1, 29.
... 70, ... 1, 2, 5, 7, 10, 14, 35, 70.
... 50, ... 1, 2, 5, 10, 25, 50.
... 80, ... 1, 2, 4, 5, 8, 10, 16 20, 40, 80.
... 66, ... 1, 2, 3, 6, 11, 22, 33, 66.
... 112, ... 1, 2, 4, 7, 8, 14, 16, 28, 56, 112.
... 120, ... 1, 2, 3, 4, 5, 6, 8, 10, 12, 15, 20, 24, 30, 40, 60, 120.
... 200, ... 1, 2, 4, 5, 8, 10, 20, 25, 40, 50, 100, 200.

Lesson II. *Common Multiples and Common Divisors.*

Teacher. Find the numbers of which 12 is a multiple.

Pupils. 12 is a multiple of 1, 2, 3, 4, and 6.

T. We will now reverse the question. Find a number which is a multiple of 1, 2, 3, 4, and 6.

P. 12 is a multiple of them.

T. For this reason 12 is said to be a *common multiple* of 1, 2, 3, 4, and 6. Find a common multiple of 3 and 5.

P. 15 is a common multiple of 3 and 5.

T. Why?

P. Because 15 is a multiple of 3, and also a multiple of 5; therefore 15 is a multiple of 3 and of 5 at the same time, or in common.

T. When is a number said to be a common multiple of two or more numbers?

P. When it contains each of these numbers exactly.

T. Find a common multiple of 5 and 4.

P. 20.

T. Find other numbers which are likewise common multiples of 5 and 4.

P. 40, 80, 100, &c.

T. But what is the least number which is a common multiple of 5 and 4?

P. 20.

T. Hence what may be said of 20?

P. 20 is the *least* common multiple of 4 and 5.

T. Find the least common multiple of

2 and 3	*Ans.* 6.
2 and 4	4.
2 and 5	10.
2 and 6	6.
2 and 7	14.
2 and 8	8.

3 and 4	*Ans.* 12.
3 and 5	15.
3 and 7	21.
3 and 8	24.
3 and 9	9.
4 and 5	20.
4 and 6	12.
4 and 7	28.
4 and 8	8.
4 and 9	36.
4 and 10..................	20.
4 and 11..................	44.
4 and 12..................	12.
2, 3, and 4	12.

Found thus:—least common multiple of

2 and 3 is evidently 6;

and the least common multiple of

6 and 4 is 12.

Hence the least common multiple of 2, 3, and 4 is 12.

Similarly, the least common multiple of

2, 3, 5	*Ans.* 30.
2, 3, 6	6.
2, 3, 7	42.

Of 2, 3, 4, 5, 6. Found thus:—

Least common multiple of 2 and 3 is 6.

... ... 6 and 4 is 12.

... ... 12 and 5 is 60.

... ... 60 and 6 is 60.

Hence least common multiple of 2, 3, 4, 5, 6, is 60.

N.B. Before proceeding, the pupils must be able to ascertain with readiness the least common multiple of, at least, any combinations of 2, 3, 4, or 5 simple numbers.

Teacher. Name a number of which 2 is a divisor?

Pupils. 2 is a divisor of 4.

T. Name another number of which 2 is likewise a divisor.

P. 6.

T. Hence, what may be said of the number 2?

P. 2 is a divisor of 4 and of 6;
2 is a *common* divisor of 4 and 6.

T. See whether the numbers 4 and 8 have a *common* divisor.

P. 2 is a common divisor of 4 and 8.

T. These numbers have another common divisor, besides 2; what is it?

P. 4.

T. Hence the common divisors of 4 and 8 are——

P. 2 and 4.

T. Find the common divisors of 6 and 12.

P. 1, 2, 3, and 6.

T. And what may be said of 6?

P. 6 is the greatest divisor of 6 and 12.

T. Find the common divisors of 12 and 20, and see which of them is the greatest.

P. The common divisors of 12 and 20 are, 1, 2, and 4;—the greatest is 4.

T. Find the greatest common divisor

Of 15 and 20 ....*Ans.*5.

6 and 24 .... 6.

12 and 30 .... 6.

4, 8, and 16, found thus:—

The greatest common divisor of 8 and 16 is 8,
and of 8 and 4 is 4.

Hence the greatest common divisor of 4, 8, and 16, is 4.

Similarly, of 5, 10, and 15 ....*Ans.*5.

9, 12, and 15 .... 3.

36, 18, and 12 .... 6.

4, and 7 .... 1.

6, 7, and 8 .... 1.

Answers to the Exercises.

1. A number is said to be a common multiple of two or more numbers, when it contains each of them exactly.
2. The least number which contains each of them exactly, is said to be the least common multiple.
3. The least common multiple

Of 2, 3, 4, 5 .... is 60.

2, 3, 4, 6 .... is 12.

2, 3, 4, 7 .... is 84.

2, 3, 4, 8 .... is 24.

2, 3, 4, 9 .... is 36.

Of 2, 3, 4, 10	is	60.
3, 4, 5, 6	is	60.
3, 4, 5, 7	is	420.
3, 4, 5, 8	is	120.
3, 4, 5, 9	is	180.
3, 4, 5, 10	is	60.
4, 5, 6, 7	is	420.
4, 5, 6, 8	is	120.
4, 5, 6, 9	is	180.
4, 5, 6, 10	is	60.
2, 4, 6, 8, 10	is	120.
2, 6, 8, 10	is	120.
3, 7, 9, 5	is	315.
5, 6, 7, 8, 9	is	2520.
2, 3, 4, 5, 6, 7, 8, 9, 10	is	2520.

4. A number which is contained in two or more numbers exactly, is said to be their common divisor.

5. And the greatest number which is so contained in two or more numbers, is said to be their greatest common divisor.

6. The greatest common divisor

Of 18 and 48	is	6.
36 and 24	is	12.
80 and 48	is	16.
60 and 84	is	12.
100 and 25	is	25.
144 and 96	is	48.

Of 91 and 104 .... is 13.
84 and 126 .... is 42.
153 and 187 .... is 17.
95 and 133 .... is 19.

Lesson III. *Prime Numbers.*

Teacher. Find the divisors of the numbers 11, 12, and 13.

Pupils. The divisors of 11 are 1 and 11.

... 12 are 1, 2, 3, 4, 6, and 12.

... 13 are 1 and 13.

T. What is there to be remarked as to the number of divisors?

P. Some numbers have more divisors than others.

T. What is the least number of divisors a number can have?

P. Two divisors; 1 and the number itself.

T. Give some instances of this.

P. The divisors of 11 are 1 and 11.

... ... 13 are 1 and 13.

... ... 7 are 1 and 7.

... ... 5 are 1 and 5.

T. Such numbers are called *prime* numbers. Find 10 numbers which are *prime* numbers.

P. 1, 2, 3, 5, 7, 11, 13, 17, 19, 23.

T. Is 14 a prime number?

P. No, because its divisors are 1, 2, 7, and 14.

T. Can a prime number be divided into two or more numbers, which, multiplied together, produce that number?

P. No; for 5 cannot be divided into two numbers, which, multiplied together, produce 5.

T. Can a number which is not prime be so divided?

P. Yes; for instance, $10=2\times5$; $12=2\times6$.

T. Name two numbers which, multiplied together, produce 20.

P. 4 and 5.

T. Are these numbers prime numbers?

P. 5 is prime, but 4 is not.

T. Since 4 is not prime, find two numbers which, multiplied together, produce 4.

P. 2 and 2.

T. [Writing on the slate.] Hence,

$$20=4\times5=2\times2\times5;$$

is this correct?

What are the numbers 2, 2, 5?

P. They are prime numbers.

T. Let us take another number which is not prime.—Choose.

P. 24.

T. Resolve 24 into two numbers.

P. 4×6.

T. Are these prime numbers?

Resolve each into two numbers.

P. $4=2\times2$, and $6=2\times3$.

T. Hence, $24=4\times6=2\times2\times2\times3$.

Resolve, in a similar manner, 20, 30, 40, into prime numbers.

P. $20=4\times5=2\times2\times5$.

$30=6\times5=2\times3\times5$.

$40=8\times5=4\times2\times5=2\times2\times2\times5$.

T. Hence, every number which is not prime, may be resolved ——

P. Into other numbers which are prime.

T. You said before, $40=8\times5$. Now, considering the numbers 8 and 5 as producing the number 40, when multiplied, they are called *factors* of 40. —Name two other factors of 40.

P. 2 and 20, or 4 and 10.

T. And if these factors be *prime* numbers, the number is said to be *resolved* into its *prime factors.* Resolve 27 into prime factors.

P. $27=3\times9=3\times3\times3$.

T. Resolve 28 into prime factors.

P. $28=4\times7=2\times2\times7$.

T. Resolve 30, 32, 36, 40, into prime factors.

P. $30=3\times10=3\times2\times5$.

$32=4\times8=4\times2\times4=2\times2\times2\times2\times2$.

$36=4\times9=2\times2\times3\times3$.

$40=4\times10=2\times2\times2\times5$.

Answers to the Exercises.

1. A number having no other divisors besides unity and the number itself, is called a prime number.

2. The prime numbers are—

1.	17.	43.	73.	101.
2.	19.	47.	79.	103.
3.	23.	53.	83.	107.
5.	29.	59.	89.	109.
7.	31.	61.	97.	113.
11.	37.	67.		119.
13.	41.	71.		

3. Numbers which, multiplied together, produce a given number, are said to be factors of that number.

4. Prime numbers which, multiplied together, produce a given number, are said to be prime factors of that number.

5. $60 = 2 \times 2 \times 3 \times 5.$

$66 = 2 \times 3 \times 11.$

$70 = 2 \times 5 \times 7.$

$74 = 2 \times 37.$

$80 = 2 \times 2 \times 2 \times 2 \times 5.$

$85 = 5 \times 17.$

89 is a prime number.

$90 = 2 \times 3 \times 3 \times 5.$

$100 = 2 \times 2 \times 5 \times 5.$

$102 = 2 \times 3 \times 17.$

$115 = 5 \times 23.$

$200 = 2 \times 2 \times 2 \times 5 \times 5.$

$300 = 2 \times 2 \times 3 \times 5 \times 5.$

$400 = 2 \times 2 \times 2 \times 2 \times 5 \times 5.$

$500 = 2 \times 2 \times 5 \times 5 \times 5.$

LESSON IV. *Square Numbers.*

This lesson supposes the pupils possessed of some elementary notions of *form*. Their attention being directed to a comparison between squares constructed upon multiple lines, they are thence to deduce the truths which follow :—

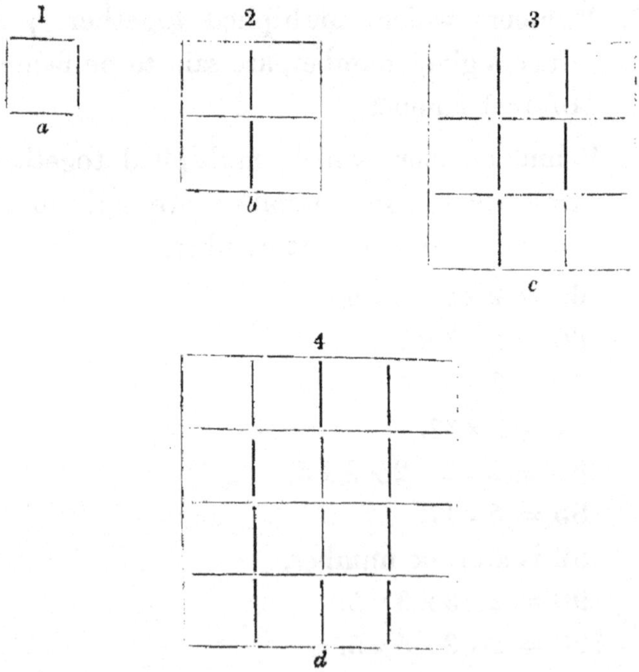

Note. These elementary notions are conveyed in "Lessons on Solids;" a little work which has long been in use in Cheam School, and is now preparing for publication. Pupils who have not received this instruction, may easily be led to the knowledge that a square is a surface having four equal sides, and its angles right angles.

If fig. *a* represent a square, of which one side be 1 inch long, each of the other sides is likewise 1 inch long, and the surface *a* is called 1 square inch; and if one side be 1 foot, 1 yard, or 1 mile long, the surface *a* is called 1 square foot, 1 square yard, or 1 square mile. Again, if fig. *b* represent a square, of which one side is 2 inches long, each side is 2 inches long; hence, dividing each side into 2 equal parts, and drawing lines joining the sections, the surface *b* is a square containing 4 squares, each equal to fig. *a;* that is, 4 square inches, 4 square feet, 4 square yards, or 4 square miles, according as fig. *a* is 1 square inch, foot, yard, or mile. Likewise, if fig. *c* represent a square, of which one side is 3 inches long, each side is 3 inches long; and hence, dividing each side into 3 equal parts, and drawing lines joining the sections, the surface *c* is a square containing 9 squares, each equal to fig. *a;* that is, 9 square inches. By a similar process, it may be ascertained, that if fig. *d* represent a square, of which one side is 4 inches long, the surface *d* is a square containing 16 square inches; and if squares be constructed upon lines 5, 6, 7, 8, 9, 10 inches long, the surfaces will be found to contain 25, 36, 49, 64, 81, and 100 square inches respectively. Hence it may be said in general:—

The square of	1 inch	length	is	1 square	inch.
...	2 inches	...	is	4 ...	inches.
...	3 ...	...	is	9 ...	...

The square of	4	inches length	is 16	square	inches.
...	5	... ...	is 25	...	...
...	6	... ...	is 36	...	...
...	7	... ...	is 49	...	...
...	8	... ...	is 64	...	...
...	9	... ...	is 81	...	...
...	10	... ...	is 100	...	...

And if instead of "inch length," we have foot, yard, or mile length, we should have square feet, square yards, square miles. Now, omitting the idea of length, it may be said generally:—

The square of	1 is 1.	The square of	6 is 36.
...	2 is 4.	...	7 is 49.
...	3 is 9.	...	8 is 64.
...	4 is 16.	...	9 is 81.
...	5 is 25.	...	10 is 100.

Originally, the numbers 1, 2, 3, ... 10, represent the length of the sides;

And the numbers .... 1, 4, 9, ... 100, the squares constructed upon them.

In the abstract the numbers 1, 2, 3, ... 10, are called *roots;*

And the numbers .... 1, 4, 9, ... 100, their *squares* respectively; and for this reason 1, 4, 9, 36, 49, &c., are called *square* numbers; and 2, 3, 5, 6, 7, &c., are *not* square numbers.

To arrive at these results, the teacher may proceed thus:—

Teacher. Tell me what you know of a square.

Pupils. A square is a surface having 4 equal sides, and 4 right angles.

T. [Describes a square upon the slate.]

1

a

If fig. *a* is a square of which one side is 1 inch long, what else is known?

P. Each of the other sides is likewise 1 inch long.

T. And what may the surface *a* be called?

P. One square inch.

T. And if one side of fig. *a* represent 1 foot, 1 yard, or 1 mile, what may the surface *a* be called each time.

P. One square foot, one square yard, or one square mile.

T. Let one of you come here, and describe a square, of which one side is two inches long.

P. [Describes upon the school slate fig. *b*.]

2

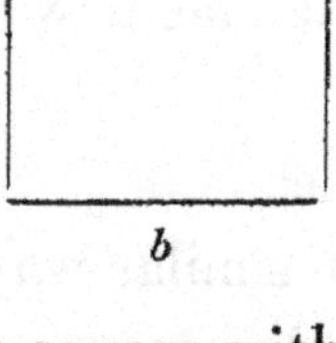

b

T. Compare this square with the former, which is the greater surface?

P. The second square, square *b*.

T. What must be done to ascertain how many times it is greater?

P. Divide each of the sides into two equal parts, and draw lines joining the sections.

T. Do it.

P. [At the slate.]

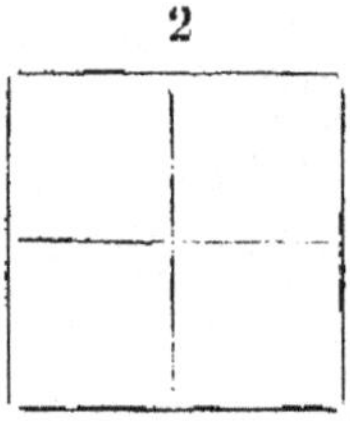

b

P. It is 4 times as great.

T. Hence if fig. *b* represent a square, of which a side is 2 inches long, the whole surface contains ——?

P. 4 square inches.

T. And if a side be 2 feet, 2 yards, or 2 miles long—

P. The whole surface contains 4 square feet, 4 square yards, or 4 square miles.

T. Describe, each on your slates, a square, of which a side is 3 inches long; proceed, as we have done just now, and tell me how many square inches it contains.

P. 9 square inches.

T. Proceed in a similar way with a square, of which a side is 4 inches long.

P. It contains 16 square inches.

T. Continue in a similar manner describing

squares whose sides are 5, 6, 7, 8, 9, and 10 inches long.

P. We have not room enough.

T. How can you remedy this?

P. We must describe smaller squares, and *suppose* the sides to be so many inches long.

T. What have you found?

P. A square, whose side is 5 inches long, contains 25 square inches;

if 6 inches long, it contains 36 square inches;

if 7 ... ... 49 ...

&c., &c., &c.

T. If a square contain 4 square inches, how many square inches lie on one side?

P. Two.

T. And how many such rows of two squares each does the whole contain?

P. Two rows.

T. If a square contain 9 square inches, how many squares lie on one side of it?

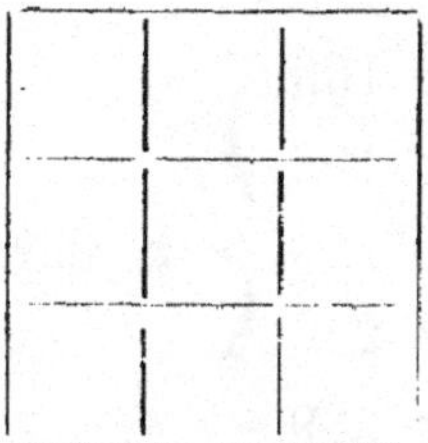

P. Three squares.

T. And how many such rows of 3 squares each does the whole contain?

P. 3 rows.

T. If a square contain 100 square inches, how many squares lie on one side of it?

P. 10 squares.

T. And how many such rows of 10 squares each does the whole contain?

P. 10 rows.

T. Now suppose a square, on one side of which lie 11 square inches, how many such rows of 11 squares each must the whole contain?

P. 11 rows.

T. How many square inches then must a square contain of which a side is 11 inches long?

P. 11 times 11, that is, 121 square inches.

T. And how many square inches does a square contain of which the length of a side is 12 inches?

P. On 1 side lie 12 squares, and there must be 12 rows each of 12 squares; it contains, therefore, 12 times 12, that is, 144 square inches.

T. How many square inches are in a square of which a side is 13, 14, 15, 16, 17, &c., inches long?

P. 13×13 or 169.
14×14 or 196.
15×15 or 225.
16×16 or 256.
17×17 or 289.
} Square inches.

T. Endeavour to express, in words, how the content (area) of a square is found, if the length of its side be known.

P. Multiply the number corresponding to the length of the side by itself.

T. Let us repeat—If the side of a square be 1 inch, 1 foot, 1 yard, 1 mile long, what is the surface called each time?

P. 1 square inch, 1 square foot, 1 square yard, 1 square mile.

T. In general, then, if the side of a square be called one, the surface must likewise be called——

P. One.

T. Or, in other words, the square of 1 is ——

P. 1.

T. In a similar way, the square of 2 is ——

P. 4.

T. The squares of 3, 4, 5, &c., are?

P. 9, 16, 25, &c.

T. Speaking of numbers, then, without reference to length, the numbers 1, 4, 9, 16, 25, &c., are called *square numbers;* and the numbers 1, 2, 3, 4, 5, &c., which represent the lengths of the sides of the squares, are then called, not *sides,* but *square roots.* What is the square of 8?

P. 64.

T. And what is the square root of 64?

P. 8.

T. What is the square root of 1?

P. 1.

T. What is the next square number, and what is its square root?

P. The next square number is 4, and its root is 2.

T. Is the square root of 2 more or less than 1?

P. It must be more than 1, but less than 2, because the square root of 4 is 2.

T. And what may be said of the square root of 3?

P. It must be more than the square root of 2, but still less than 2.

T. The next number is 4, and its square root is exactly ——

P. 2.

T. The next number is 5, and its square root is ——

P. More than 2, but less than 3, because the square root of 9 is 3.

T. The next numbers are 6, 7, 8, and their square roots are ——

P. All more than 2, but still less than 3.

T. What are the square roots of 10, 11, 12, 13, 14, 15?

P. All more than 3, but less than 4.

T. What is the square root of 50?

P. More than 7, but less than 8.

T. Now, if a number is more than 7, but less than 8, how much must it be?

P. A little more than 7;—7 and a part of 1.

T. You do not know what part of 1 to call this.

In our next lesson we will speak of the parts of 1, or *fractions,* as they are called.

The ideas of the division of unity, and of the necessity of this division, are thus strongly awakened in the minds of the pupils; and the moment has now arrived to acquaint them with the parts of unity, or what are called fractions.

Answers to the Exercises.

Ans. 1. 1, 4, 9, 16, 25, 36, 49, 64, 81, 100, 121, 144, 169, 196.

2. 100, 400, 900, 1600, 2500, 3600, 4900, 6400, 8100, 10000.

3. The square root of a number represents the side of which the number represents the square; or the square root of a number is such as, being multiplied by itself, produces the given number.

4. More than 9, but less than 10.

CHAPTER VI.—FRACTIONS.

§ 1. DENOMINATIONS OF HALVES, THIRDS, FOURTHS, &c.

LESSON I. *Halves.*

THE first notions of fractions are obtained, like the first ideas of number, from the senses. As before, a line furnishes sufficient means to effect that purpose. To obtain halves, then, a line is divided into 2 equal parts; to obtain the idea of thirds, it is divided into 3 equal parts; to obtain fourths, fifths, sixths, &c., it is divided into 4, 5, 6, &c., equal parts successively, as will be seen in detail from what follows.

Teacher. Draw a straight line on your slate, and divide it into two equal parts.

Pupils. ——|——.

T. What part is one of these divisions of the whole line?

P. One-half.

T. What is the other part of the whole line?

P. Likewise one-half.

T. What are the two parts taken together?

P. The whole line.

T. How many halves, then, are in one line?

P. Two halves.

T. By what means is the half of any thing obtained?

P. By dividing it into two equal parts, and taking one of these parts.

T. Imagine the number 1 divided into 2 equal parts; what will you call each of these parts?

P. One-half.

T. One-half is written thus, $\frac{1}{2}$. How many halves together make 1?

P. 2 halves.

T. If we wished to express this shortly in writing, how could it be done?

P. Thus, $\frac{1}{2}+\frac{1}{2}=1$.

T. Two halves is written thus, $\frac{2}{2}$; hence 1 is the same as ——. Write the answer on your slates.

P. $1 = \frac{2}{2}$.

T. Since in 1 there are $\frac{2}{2}$, how many halves are there in 2?

P. Twice $\frac{2}{2}$; that is, 4 halves.

T. How will you write 4 halves?

P. Thus, $\frac{4}{2}$.

T. Hence 2 is the same as $\frac{4}{2}$;—express this shortly in writing.

P. $\frac{4}{2} = 2$.

T. And how much is $\frac{3}{2}$?

P. One and one-half.

T. Write this on your slates.

P. $\frac{3}{2} = 1\frac{1}{2}$.

T. How many halves are there in 3, 4, 5, 6, 7, &c.?

P. 6, 8, 10, 12, 14, &c., halves.

T. You say 14 halves are 7; tell me why?

P. Because in 1 there are $\frac{2}{2}$, and, therefore, in 7 there must be 7 times $\frac{2}{2}$, that is, $\frac{14}{2}$.

T. How many halves are there in 17?

P. In 1 there are $\frac{2}{2}$, and in 17 there are 17 times $\frac{2}{2}$, which are $\frac{34}{2}$.

T. Let us write what we have learnt upon the slate. [Writing.]

$1 = \frac{2}{2}$.	$6 = \frac{12}{2}$.
$2 = \frac{4}{2}$.	$7 = \frac{14}{2}$.
$3 = \frac{6}{2}$.	$8 = \frac{16}{2}$.
$4 = \frac{8}{2}$.	$9 = \frac{18}{2}$.
$5 = \frac{10}{2}$.	$10 = \frac{20}{2}$.

&c.

Read this.

P. [Read] 1 is equal to 2 halves.
2 are equal to 4 halves.
3 are equal to 6 halves,
&c., &c.

T. We will now reverse the questions.
How much are $\frac{2}{2}$, $\frac{4}{2}$, $\frac{6}{2}$, $\frac{8}{2}$, &c.

P. 1, 2, 3, 4, &c.

T. How much are $\frac{48}{2}$?

P. 24.

T. Tell me how you found this.

P. Because $\frac{2}{2} = 1$, as often as $\frac{2}{2}$ are contained in $\frac{48}{2}$, so many ones we shall have. Now $\frac{2}{2}$ are contained in $\frac{48}{2}$ 24 times; therefore $\frac{48}{2} = 24$.

T. Instead of saying *ones*, it would be better to say—

P. Units.

T. Hence $\frac{48}{2}$ are equal to—

P. 24 units.

T. How many units are there in $\frac{17}{2}$, $\frac{37}{2}$, $\frac{58}{2}$, $\frac{97}{2}$?

P. $8\frac{1}{2}$, $18\frac{1}{2}$, 29, $48\frac{1}{2}$.

N.B. The pupils must be able to state how they have obtained these answers.

T. There is another sort of questions connected with the idea of halves; when you are able to answer them, we shall proceed. It is this:—

What is $\frac{1}{2}$ of 1?

P. $\frac{1}{2}$.

T. What is $\frac{1}{2}$ of 2?

P. $\frac{2}{2}$ or 1.

T. What is $\frac{1}{2}$ of 3?

P. $\frac{3}{2}$, or $1\frac{1}{2}$.

T. What is $\frac{1}{2}$ of 4, of 5, of 6, of 7, &c.?

P. 2, $2\frac{1}{2}$, 3, $3\frac{1}{2}$, &c.

T. What is $\frac{1}{2}$ of 17?

P. Because $\frac{1}{2}$ of 1 is $\frac{1}{2}$.

$\therefore$ $\frac{1}{2}$ of 17 must be $17 \times \frac{1}{2}$;

that is, $\frac{17}{2}$, or $8\frac{1}{2}$.

T. What are $\frac{2}{2}$ of a number?

P. The whole number.

T. What are $\frac{2}{2}$ of 59?

P. 59.

T. And if one-half of a number be taken 3 times, what must the answer be?

P. More than the number; the number more one-half of the number.

T. What then are $\frac{3}{2}$ of 12?

P. 18; because $\frac{1}{2}$ of $12 = 6$, therefore $\frac{3}{2}$ of 12 $= 3 \times 6 = 18$.

or because $\frac{2}{2}$ of $12 = 12$, and $\frac{1}{2}$ of $12 = 6$, therefore $\frac{3}{2}$ of $12 = 12 + 6 = 18$.

T. And if one-half of a number be taken 4 times, what must be the answer?

P. Twice the number; because $\frac{2}{2}$ of a number are equal to the number, and $\frac{4}{2}$ of a number must therefore be equal to twice the number.

T. How much is $\frac{4}{2}$ of 8?

P. 16; because $\frac{1}{2}$ of $8 = 4$, therefore, $\frac{4}{2}$ of $8 = 4 \times 4 = 16$;

or,

because $\frac{2}{2}$ of $8 = 8$;

therefore $\frac{4}{2}$ of $8 = 2 \times 8 = 16$.

T. What are $\frac{5}{2}$ of 20?

P. 50.

T. What are $\frac{6}{2}$ of 4?

P. 3×4 or 12.

&c.

T. Before we proceed, I wish you to tell me

what sort of questions we have had since we began fractions.

P. 1. The first sort was,

How many halves are there in

1, 2, 3, 4, 5, &c.

2. The second sort was,

How many units are there in

$\frac{2}{2}$, $\frac{3}{2}$, $\frac{4}{2}$, &c., $\frac{17}{2}$, &c.

3. The third sort was,

How much is $\frac{3}{2}$ of 14, $\frac{5}{2}$ of 26, &c.

The teacher and pupils are to give questions to the class, after which the exercises in Part II are to be taken up.

Answers to the Exercises.

Ans. 1. If a number be divided into two equal parts, each of those parts is called one half of the number.

2. $\frac{154}{2}$; $\frac{190}{2}$; $\frac{226}{2}$; $\frac{690}{2}$; $\frac{1578}{2}$; $\frac{1972}{2}$.

3. $38\frac{1}{2}$; $47\frac{1}{2}$; $56\frac{1}{2}$; $172\frac{1}{2}$; $394\frac{1}{2}$; 493.

4. $\frac{3}{2}$ of $49 = 73\frac{1}{2}$; $\frac{11}{2}$ of $12 = 66$;
$\frac{5}{2}$ of $58 = 145$; $\frac{13}{2}$ of $5 = 32\frac{1}{2}$;
$\frac{7}{2}$ of $89 = 311\frac{1}{2}$; $\frac{14}{2}$ of $3 = 21$;
$\frac{8}{2}$ of $100 = 400$; $\frac{15}{2}$ of $20 = 150$;
$\frac{9}{2}$ of $43 = 193\frac{1}{2}$; $\frac{19}{2}$ of $15 = 142\frac{1}{2}$.

Lesson II. *Thirds.*

Teacher. [Draws a straight line upon the slate.]

a——|——|——*b*. Let us divide the straight line
c *d*

a b into 3 equal parts, and put the letters *c d* at the points of section. What part of the whole line is *a c*?

P. One-third.

T. And *c d*, *d b*?

P. Each one-third.

T. How many thirds then make up the whole line?

P. 3 thirds.

T. What part of the whole line is *a d*?

P. 2 thirds.

T. By what means is a third of any thing obtained?

P. By dividing it into three equal parts, and taking one of them.

T. And by what means are two-thirds of any thing obtained?

P. By dividing it into 3 equal parts, and taking 2 of them.

T. If now we imagine the number 1 divided into 3 equal parts, what will you call each of the parts?

P. One-third.

T. One-third is written thus $\frac{1}{3}$. Let us examine this mode of writing one-third. What does the number 3 express?

P. It shows that 1 has been divided into 3 equal parts.

T. And what does the 1 express?

P. That 1 of these 3 equal parts has been taken.

T. How are two-thirds of one obtained?

P. By supposing 1 divided into 3 equal parts, and taking two of them.

T. Try, now, to express two-thirds in writing.

P. $\frac{2}{3}$.

T. You have written 2 above 3, and separated both numbers by a small line drawn between them. Some of you have put 3 above 2. Did I proceed thus in writing $\frac{1}{3}$?—Which number have I put below the line?—which above? Remember, then, the number which shows into how many parts 1 has been divided, is put below that which shows how many of the parts are taken. Hence, what does $\frac{3}{2}$ mean?

P. That 1 has been divided into 2 equal parts, and that 3 of such parts have been taken. It means $\frac{1}{2}$ taken three times.

T. Now express 3 thirds.

P. $\frac{3}{3}$.

T. And what are $\frac{3}{3}$ equal to?

P. $\frac{3}{3} = 1$.

T. In our last lesson on halves, we had $\frac{2}{2} = 1$; what were the questions after that?

P. How many halves are in 2, 3, 4, &c.

T. Let us, then, ask ourselves how many thirds there are in 2, 3, 4, 5, &c.

P. $\frac{6}{3}$, $\frac{9}{3}$, $\frac{12}{3}$, $\frac{15}{3}$, &c.

T. How many thirds are there in 21?

P. $\frac{63}{3}$; because in 1 there are $\frac{3}{3}$; in 21, therefore, there must be 21 times $\frac{3}{3}$, that is $\frac{63}{3}$.

T. What was the next sort of questions in our lessons on halves?

P. Just the reverse. How many units are there in $\frac{4}{3}$, $\frac{5}{3}$, $\frac{6}{3}$ .... $\frac{100}{3}$, &c.

T. How many units are there in $\frac{49}{3}$?

P. $16\frac{1}{3}$; because $\frac{3}{3} = 1$; therefore, as often as $\frac{3}{3}$ are contained in $\frac{49}{3}$, so many units there will be. Now $\frac{3}{3}$ are contained in $\frac{49}{3}$ 16 times; and $\frac{1}{3}$ more; therefore $\frac{49}{3} = 16\frac{1}{3}$.

T. And what was the third sort of questions in halves.

P. How much is $\frac{3}{2}$ of 18; $\frac{4}{2}$ of 29, &c.

T. How much, then, is $\frac{1}{3}$ of 2?

P. $\frac{2}{3}$; because $\frac{1}{3}$ of $1 = \frac{1}{3}$; therefore $\frac{1}{3}$ of 2 must be 2 times $\frac{1}{3}$, that is $\frac{2}{3}$.

T. How much is $\frac{1}{3}$ of 4, 5, 6, &c., 100?

P. $\frac{1}{3}$ of $4 = 1\frac{1}{3}$;
$\frac{1}{3}$ of $5 = 1\frac{2}{3}$;
$\frac{1}{3}$ of $6 = 2$; &c.
and $\frac{1}{3}$ of $100 = 33\frac{1}{3}$.

T. How much is $\frac{2}{3}$ of 2?

P. $1\frac{1}{3}$; because $\frac{1}{3}$ of $2 = \frac{2}{3}$; therefore $\frac{2}{3}$ of 2 must be twice $\frac{2}{3}$; that is $\frac{4}{3}$, or $1\frac{1}{3}$.

T. How much is $\frac{2}{3}$ of 7, 14, 20, 50.

P. Because $\frac{1}{3}$ of $7 = \frac{7}{3} = 2\frac{1}{3}$;
therefore $\frac{2}{3}$ of $7 = 2 \times 2\frac{1}{3} = 4\frac{2}{3}$.

Again,

Because $\frac{1}{3}$ of 14 $= \frac{14}{3} = 4\frac{2}{3}$;

therefore $\frac{2}{3}$ of 14 $= 2 \times 4\frac{2}{3} = 9\frac{1}{3}$.

Also,

Because $\frac{1}{3}$ of 20 $= \frac{20}{3} = 6\frac{2}{3}$;

therefore $\frac{2}{3}$ of 20 $= 2 \times 6\frac{2}{3} = 13\frac{1}{3}$.

And,

Because $\frac{1}{3}$ of 50 $= \frac{50}{3} = 16\frac{2}{3}$;

therefore $\frac{2}{3}$ of 50 $= 2 \times 16\frac{2}{3} = 33\frac{1}{3}$.

Similar solutions are required for each of the questions given by the teacher and the pupils, after which the exercises in Part II are taken up.

Answers to the Exercises.

Ans. 1. $\frac{5}{2}$ means that 1 has been divided into 2 equal parts, and that one of these parts has been taken 5 times.

And $\frac{5}{3}$ means that 1 has been divided into 3 equal parts, and that one of these parts has been taken 5 times.

2. $7\frac{1}{3} = \frac{22}{3}$; $19\frac{2}{3} = \frac{59}{3}$.

$8\frac{2}{3} = \frac{26}{3}$; $87\frac{1}{3} = \frac{262}{3}$.

3. $\frac{28}{3} = 9\frac{1}{3}$, and $\frac{28}{2} = 14$.

$\frac{45}{3} = 15$, and $\frac{45}{2} = 22\frac{1}{2}$.

$\frac{117}{3} = 39$, and $\frac{117}{2} = 58\frac{1}{2}$.

$\frac{141}{3} = 47$, and $\frac{141}{2} = 70\frac{1}{2}$.

$\frac{201}{3} = 67$, and $\frac{201}{2} = 100\frac{1}{2}$.

Ans. 4. $\frac{2}{3}$ of 19 = $12\frac{2}{3}$; $\frac{4}{3}$ of 59 = $78\frac{2}{3}$.
$\frac{2}{3}$ of 53 = $35\frac{1}{3}$; $\frac{5}{3}$ of 48 = 80.
$\frac{2}{3}$ of 80 = $53\frac{1}{3}$; $\frac{6}{3}$ of 50 = 100.
$\frac{2}{3}$ of 86 = $57\frac{1}{3}$; $\frac{7}{3}$ of 81 = 189.
$\frac{2}{3}$ of 100 = $66\frac{2}{3}$; $\frac{8}{3}$ of 82 = $218\frac{2}{3}$.
$\frac{2}{3}$ of 126 = 84; $\frac{9}{3}$ of 100 = 300.

LESSON III. *Fourths.*

Teacher. Draw a straight line on your slates and divide it into 4 equal parts.

Pupils. ——|——|——|——.

T. What part of the whole line is one of these parts?

P. 1 fourth.

T. And what are 2 parts?

P. 2 fourths.

T. What are 3 parts?

P. 3 fourths.

T. And what are the 4 parts taken together?

P. 4 fourths, or the whole line.

T. If, now, we imagine the number 1 to be divided into 4 equal parts, what is each of the parts called?

P. One-fourth.

T. Write in figures one-fourth.

P. $\frac{1}{4}$.

T. What are 2 parts called?

P. $\frac{2}{4}$.

T. What are 3 parts called?

P. $\frac{3}{4}$.

T. What does the number 4 indicate?

P. That 1 has been divided into 4 equal parts.

T. It is called the *denominator.* Thus, in the fraction $\frac{2}{3}$, what is the name of the parts into which 1 has been divided?

P. Thirds.

T. Hence the number 3 names or denominates the parts into which 1 has been divided. And in the fraction $\frac{3}{4}$, what does 3 indicate?

P. That 3 of the 4 equal parts have been taken.

T. That is, it shows the number of parts taken, and it is thence called the *numerator.* In $\frac{2}{3}$, which is the denominator—which the numerator?

P. 3 is the denominator, and 2 the numerator.

T. How many fourths are there in 1, 2, 3, 4, 5, &c......100?

P. $\frac{4}{4}$, $\frac{8}{4}$, $\frac{12}{4}$, $\frac{16}{4}$, $\frac{20}{4}$......$\frac{400}{4}$.

T. Our second sort of questions are the reverse of the above; that is, how many units are there in $\frac{5}{4}$, $\frac{6}{4}$, $\frac{7}{4}$, $\frac{8}{4}$, $\frac{9}{4}$, &c.......$\frac{100}{4}$?

P. $1\frac{1}{4}$, $1\frac{2}{4}$, $1\frac{3}{4}$, 2, $2\frac{1}{4}$......25.

T. And our third sort of questions are, what is $\frac{1}{4}$ of 2, 3, 4......17?

P. $\frac{2}{4}$, $\frac{3}{4}$, 1......$4\frac{1}{4}$.

T. What is $\frac{2}{4}$ of 37?

P. $\frac{1}{4}$ of $37=\frac{37}{4}=9\frac{1}{4}$; and

$\frac{2}{4}$ of $37=2\times 9\frac{1}{4}=18\frac{2}{4}$.

T. How much is $\frac{3}{4}$ of 23?

P. $\frac{1}{4}$ of $23 = \frac{23}{4} = 5\frac{3}{4}$; and
$\frac{3}{4}$ of $23 = 3 \times 5\frac{3}{4} = 17\frac{1}{4}$.

Answers to the Exercises.

Ans. 1. The denominator of a fraction indicates how many equal parts 1 has been divided into; and the numerator shows how many of the equal parts have been taken.

2. $5\frac{3}{4} = \frac{23}{4}$; $7\frac{3}{4} = \frac{31}{4}$; $19\frac{1}{4} = \frac{77}{4}$; $37\frac{3}{4} = \frac{151}{4}$; $113\frac{1}{4} = \frac{453}{4}$; $239\frac{3}{4} = \frac{959}{4}$; $271\frac{1}{4} = \frac{1085}{4}$.

3. $\frac{17}{2} = 8\frac{1}{2}$; $\frac{17}{3} = 5\frac{2}{3}$; $\frac{17}{4} = 4\frac{1}{4}$.
$\frac{39}{2} = 19\frac{1}{2}$; $\frac{53}{3} = 17\frac{2}{3}$; $\frac{39}{4} = 9\frac{3}{4}$.
$\frac{86}{2} = 43$; $\frac{86}{3} = 28\frac{2}{3}$; $\frac{86}{4} = 21\frac{2}{4}$.
$\frac{153}{2} = 76\frac{1}{2}$; $\frac{153}{3} = 51$; $\frac{153}{4} = 38\frac{1}{4}$.
$\frac{217}{3} = 72\frac{1}{3}$; $\frac{217}{4} = 54\frac{1}{4}$; $\frac{586}{3} = 195\frac{1}{3}$;
$\frac{586}{4} = 146\frac{2}{4}$; $\frac{666}{3} = 222$; $\frac{666}{4} = 166\frac{2}{4}$.

4. $\frac{2}{3}$ of $17 = 11\frac{1}{3}$; $\frac{6}{4}$ of $33 = 49\frac{2}{4}$.
$\frac{3}{4}$ of $17 = 12\frac{3}{4}$; $\frac{7}{4}$ of $25 = 43\frac{3}{4}$.
$\frac{2}{3}$ of $82 = 54\frac{2}{3}$; $\frac{8}{4}$ of $136 = 272$.
$\frac{3}{4}$ of $82 = 61\frac{2}{4}$; $\frac{9}{4}$ of $116 = 261$.
$\frac{3}{4}$ of $96 = 72$; $\frac{10}{4}$ of $44 = 110$.
$\frac{5}{4}$ of $71 = 88\frac{3}{4}$; $\frac{11}{4}$ of $29 = 79\frac{3}{4}$.

LESSON IV. *Fifths, Sixths, Sevenths.*

If a straight line be divided into 5, 6, 7 equal parts, the pupils will arrive at the idea of 5ths, 6ths, 7ths. A process quite analogous to that pursued in the previous lessons, is then applied to the number 1. The following are instances of these sorts of questions, referred to in Lessons I, II, III.

1. Reduce 1, 2, 3, 4......98, 99, 100 to 5ths.
   ... 1, 2, 3, 4......... 99, 100 to 6ths.
   ... 1, 2, 3, 4......... 99, 100 to 7ths.

2. Reduce to units $\frac{6}{5}$, $\frac{7}{5}$, $\frac{8}{15}$...... $\frac{99}{5}$, $\frac{100}{5}$.
   ... ... $\frac{7}{6}$, $\frac{8}{6}$, $\frac{9}{6}$ ...... $\frac{99}{6}$, $\frac{100}{6}$.
   ... ... $\frac{8}{7}$, $\frac{9}{7}$, $\frac{10}{7}$...... $\frac{99}{7}$, $\frac{100}{7}$.

3. How much is $\frac{4}{5}$ of 27? $\frac{3}{5}$ of 39; $\frac{2}{5}$ of 63, &c.?
   ... ... $\frac{2}{6}$ of 31; $\frac{3}{6}$ of 48; $\frac{4}{6}$ of 55, &c.?
   ... ... $\frac{2}{7}$ of 17; $\frac{3}{7}$ of 30; $\frac{4}{7}$ of 43; $\frac{5}{7}$ of 100, &c.?

Answers to the Exercises.

Ans. 1. $\frac{215}{5}$; $\frac{225}{5}$; $\frac{585}{5}$; $\frac{695}{5}$; $\frac{1090}{5}$; $\frac{1595}{5}$.

2. $17\frac{2}{5}$; $10\frac{3}{5}$; $19\frac{2}{5}$; $22\frac{1}{5}$; $33\frac{4}{5}$; $43\frac{2}{5}$; $68\frac{3}{5}$; $97\frac{1}{5}$; $112\frac{1}{5}$; $135\frac{4}{5}$.

Ans. 3. $\frac{4}{5}$ of 53 = $42\frac{2}{5}$; $\frac{3}{5}$ of 216 = $129\frac{3}{5}$;
$\frac{3}{5}$ of 89 = $53\frac{2}{5}$; $\frac{2}{5}$ of 331 = $132\frac{2}{5}$;
$\frac{2}{5}$ of 117 = $46\frac{4}{5}$; $\frac{4}{5}$ of 459 = $367\frac{1}{5}$;
$\frac{4}{5}$ of 189 = $151\frac{1}{5}$.

4. $\frac{534}{6}$; $\frac{582}{6}$; $\frac{690}{6}$; $\frac{1074}{6}$; $\frac{1428}{6}$; $\frac{2754}{6}$.

5. $15\frac{4}{6}$; $24\frac{5}{6}$; $30\frac{3}{6}$; $36\frac{1}{6}$; $57\frac{1}{6}$; 76; 96; 113; $148\frac{2}{6}$.

6. $\frac{1}{6}$ of 86 = $14\frac{2}{6}$; $\frac{5}{6}$ of 187 = $155\frac{5}{6}$;
$\frac{2}{6}$ of 93 = 31; $\frac{7}{6}$ of 44 = $51\frac{2}{6}$;
$\frac{3}{6}$ of 115 = $57\frac{3}{6}$; $\frac{8}{6}$ of 80 = $106\frac{4}{6}$;
$\frac{4}{6}$ of 139 = $92\frac{4}{6}$.

7. $\frac{651}{7}$; $\frac{819}{7}$; $\frac{1302}{7}$; $\frac{1498}{7}$; $\frac{2128}{7}$; $\frac{2005}{7}$.

8. $7\frac{4}{7}$; $12\frac{2}{7}$; $13\frac{3}{7}$; $21\frac{6}{7}$; 27; 31; $35\frac{4}{7}$; $45\frac{2}{7}$; $59\frac{2}{7}$; 77; $97\frac{1}{7}$.

9. $\frac{1}{7}$ of 86 = $12\frac{2}{7}$; $\frac{6}{7}$ of 200 = $171\frac{3}{7}$;
$\frac{2}{7}$ of 94 = $26\frac{6}{7}$; $\frac{6}{7}$ of 319 = $273\frac{3}{7}$;
$\frac{3}{7}$ of 112 = 48; $\frac{5}{7}$ of 480 = $342\frac{6}{7}$;
$\frac{4}{7}$ of 186 = $106\frac{2}{7}$; $\frac{4}{7}$ of 590 = $337\frac{1}{7}$;
$\frac{5}{7}$ of 190 = $135\frac{5}{7}$; $\frac{3}{7}$ of 640 = $274\frac{2}{7}$.

Lesson V. *Eighths, Ninths, Tenths.*

Answers to the Exercises.

Ans. 1. $\frac{472}{8}$; $\frac{688}{8}$; $\frac{776}{8}$; $\frac{1048}{8}$; $\frac{1416}{8}$; $\frac{1744}{8}$; $\frac{2552}{8}$.

2. $11\frac{5}{8}$; 18; $29\frac{5}{8}$; $39\frac{5}{8}$; $56\frac{5}{8}$; $64\frac{4}{8}$; 84; $97\frac{1}{8}$; 106; $112\frac{4}{8}$.

3. $\frac{1}{8}$ of 49 = $6\frac{1}{8}$; $\frac{5}{8}$ of 156 = $97\frac{4}{8}$;

$\frac{2}{8}$ of 73 = $18\frac{2}{8}$; $\frac{6}{8}$ of 200 = 150;

$\frac{3}{8}$ of 100 = $37\frac{4}{8}$; $\frac{7}{8}$ of 346 = $302\frac{6}{8}$;

$\frac{4}{8}$ of 115 = $57\frac{4}{8}$.

4. $\frac{774}{9}$; $\frac{945}{9}$; $\frac{1377}{9}$; $\frac{1944}{9}$; $\frac{3429}{9}$.

5. $4\frac{7}{9}$; $12\frac{4}{9}$; $20\frac{6}{9}$; $24\frac{1}{9}$; $39\frac{7}{9}$.

6. $\frac{1}{9}$ of 86 = $9\frac{5}{9}$; $\frac{5}{9}$ of 134 = $74\frac{4}{9}$;

$\frac{2}{9}$ of 144 = 32; $\frac{6}{9}$ of 242 = $161\frac{3}{9}$;

$\frac{3}{9}$ of 63 = 21; $\frac{7}{9}$ of 313 = $243\frac{4}{9}$;

$\frac{4}{9}$ of 115 = $51\frac{1}{9}$; $\frac{8}{9}$ of 516 = $458\frac{6}{9}$.

7. $\frac{860}{10}$; $\frac{1440}{10}$; $\frac{2170}{10}$; $\frac{3920}{10}$.

8. $9\frac{3}{10}$; $14\frac{3}{10}$; $18\frac{7}{10}$; $25\frac{9}{10}$; $37\frac{7}{10}$.

Ans. 9. $\frac{1}{10}$ of 87 = $8\frac{7}{10}$;

$\frac{2}{10}$ of 115 = 23;

$\frac{3}{10}$ of 186 = $55\frac{8}{10}$;

$\frac{4}{10}$ of 217 = $86\frac{8}{10}$;

$\frac{5}{10}$ of 316 = 158;

$\frac{6}{10}$ of 419 = $251\frac{4}{10}$;

$\frac{7}{10}$ of 528 = $369\frac{6}{10}$;

$\frac{8}{10}$ of 674 = $539\frac{2}{10}$;

$\frac{9}{10}$ of 749 = $674\frac{1}{10}$.

§ 2. OF THE DIFFERENT EXPRESSIONS FOR THE SAME FRACTIONS.

Lesson I.

Different Expressions for Halves.

If a straight line be divided into 2 equal parts, and each of these parts be further subdivided into 2, 3, 4, 5, &c., equal parts, the whole line will thus be divided into 4, 6, 8, 10, &c., parts, of which the half contains 2, 3, 4, 5, &c.

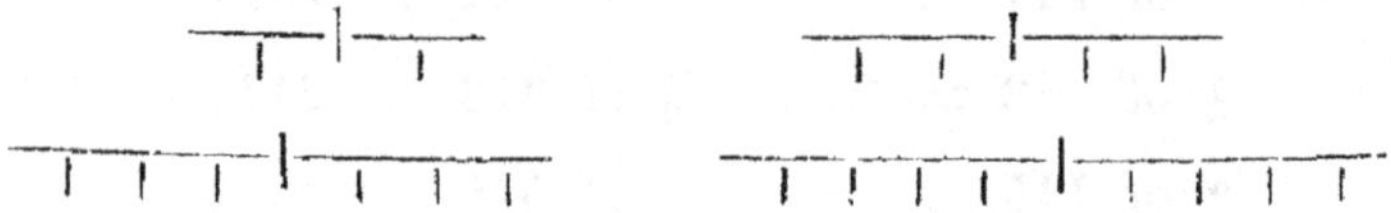

That is, one half of the line is the same as 2 fourths, as 3 sixths, as 4 eighths, as 5 tenths, &c. If the same kind of division be afterwards applied to the number 1, we shall have

$\frac{1}{2} = \frac{2}{4} = \frac{3}{6} = \frac{4}{8} = \frac{5}{10} = \frac{6}{12}$, &c.

By a similar process it will be found

That $\frac{1}{3} = \frac{2}{6} = \frac{3}{9} = \frac{4}{12} = \frac{5}{15}$, &c.

And so on as will be shown in the following lessons ;—

Teacher. Draw a straight line on your slates, which divide into 2 equal parts. What is each of the parts called?

Pupils. One-half.

T. Subdivide each of these halves again into 2 equal parts, into how many parts is the whole line then divided?

P. Into 4 equal parts.

T. What then is each of them called?

P. One-fourth.

T. And how many of these fourths are there in half the line?

P. Two.

T. Hence it may be said that one half is the same as ——

P. One half is the same as two-fourths.

T. Instead of dividing each half into 2 equal parts, divide it into 3 equal parts, and state what you observe.

P. One half is the same as 3 sixths.

T. Continue to divide each half into 4, 5, 6, &c., equal parts, and state each time what you observe.

P. One half is the same as 4 eighths, or as 5 tenths, or as 6 twelfths, &c.

T. Let us now apply what we have learnt to the number 1. Suppose 1 to be divided into 2 equal parts, each of them then is called ——

P. $\frac{1}{2}$.

T. And if each of these halves be divided into 2 equal parts, the number 1 is then divided into ——

P. Fourths.

T. And $\frac{1}{2}$ contains how many of these fourths?

P. Two.

T. Come here and write this on the slate.

P. $\frac{1}{2} = \frac{2}{4}$.

T. And if each $\frac{1}{2}$ is divided into 3 equal parts, the number 1 is then divided into ——

P. 6 sixths.

T. And $\frac{1}{2}$ contains how many of these 6 sixths?

P. Three.

T. Hence $\frac{1}{2} = \frac{2}{4} = \frac{3}{6}$. In how many different ways have we now expressed $\frac{1}{2}$?

P. In two different ways.

T. Try to find 20 different expressions for $\frac{1}{2}$.

P. $\frac{1}{2} = \frac{2}{4} = \frac{3}{6} = \frac{4}{8} = \frac{5}{10} = \frac{6}{12}, \frac{7}{14}$, &c.

T. Express $\frac{1}{2}$ in 14ths.

P. $1 = \frac{14}{14}$; and $\frac{1}{2}$ therefore $=$ $\frac{1}{2}$ of $\frac{14}{14}$, which is $\frac{7}{14}$.

T. In which numbers can $\frac{1}{2}$ be expressed?

P. In 4ths, 6ths, 8ths, 10ths, 12ths, &c.

T. Generally speaking, in which numbers?

P. In all numbers which are multiples of 2.

T. Can $\frac{1}{2}$ be expressed in 3ds, 5ths, or 7ths?

P. No, because 3, 5, 7, are not multiples of 2.

LESSON II. *Different Expressions for Thirds, Fourths, Fifths, &c.*

Teacher. Divide a line into 3 equal parts, what is each of the parts called?

Pupils. One-third.

T. Subdivide each of these 3ds into 2 equal parts, and state what you observe.

P. One-third is the same as 2-sixths.

T. If we wish to find other different expressions for $\frac{1}{3}$, what must be done?

P. Each third must be divided successively into 3, 4, 5, 6, &c., equal parts.

T. Apply the same way of reasoning to the number 1, and find 6 different expressions for $\frac{1}{3}$.

P. $\frac{1}{3} = \frac{2}{6} = \frac{3}{9} = \frac{4}{12} = \frac{5}{15} = \frac{6}{18}$.

T. In what numbers can $\frac{1}{3}$ be expressed?

P. In 6ths, 9ths, 12ths, &c.; in general, in all numbers which are multiples of 3.

T. Instead of saying, "express $\frac{1}{3}$ in 27ths," it is usual to say, "*reduce* $\frac{1}{3}$ to 27ths;" that is, how many 27ths are equal to $\frac{1}{3}$?

P. $\frac{1}{3} = \frac{9}{27}$.

T. Why?

P. Because $1 = \frac{27}{27}$, $\frac{1}{3}$ therefore must be $\frac{1}{3}$ of $\frac{27}{27}$, which are $\frac{9}{27}$.

T. And how many 27ths are equal to $\frac{2}{3}$, or reduce $\frac{2}{3}$ to 27ths?

P. $\frac{2}{3} = \frac{18}{27}$; because $\frac{1}{3} = \frac{9}{27}$, $\frac{2}{3}$ must be twice $\frac{9}{27}$, which are $\frac{18}{27}$.

T. Reduce $\frac{2}{3}$ to 36ths.

P. $1 = \frac{36}{36}$; $\frac{1}{3}$ therefore must be $\frac{1}{3}$ of $\frac{36}{36}$, which is $\frac{12}{36}$; and $\frac{2}{3}$ must be twice $\frac{12}{36}$, which is $\frac{24}{36}$.

T. Hence if you know in what numbers 4ths, 5ths, 6ths, &c., can be expressed, you will be able to find different expressions for $\frac{3}{4}$, $\frac{4}{5}$, $\frac{5}{6}$, &c. In what numbers can 4ths be expressed?

P. In 8ths, 12ths, 16ths, &c.; in all numbers which are multiples of 4.

T. Reduce $\frac{1}{4}$ to 20ths.

P. $\frac{1}{4} = \frac{5}{20}$, because $1 = \frac{20}{20}$; therefore $\frac{1}{4}$ must be $\frac{1}{4}$ of $\frac{20}{20}$, which are $\frac{5}{20}$.

T. Reduce $\frac{3}{4}$ to 28ths.

P. $\frac{3}{4} = \frac{21}{28}$, because $1 = \frac{28}{28}$;
therefore $\frac{1}{4} = \frac{1}{4}$ of $\frac{28}{28} = \frac{7}{28}$, and
$\frac{3}{4} = 3 \times \frac{7}{28} = \frac{21}{28}$.

T. Find 6 different expressions for $\frac{3}{4}$.

P. $\frac{3}{4} = \frac{6}{8} = \frac{9}{12} = \frac{12}{16} = \frac{15}{20} = \frac{18}{24} = \frac{21}{28}$.

T. In what numbers can 5ths be expressed?

P. In 10ths, 15ths, 20ths, &c.; in all numbers which are multiples of 5.

T. Reduce $\frac{1}{5}$ to 25ths.

P. $\frac{1}{5} = \frac{5}{25}$, because $1 = \frac{25}{25}$; therefore $\frac{1}{5} = \frac{1}{5}$ of $\frac{25}{25} = \frac{5}{25}$.

T. Reduce $\frac{4}{5}$ to 60ths.

P. $\frac{4}{5} = \frac{48}{60}$, because $1 = \frac{60}{60}$;
therefore $\frac{1}{5} = \frac{1}{5}$ of $\frac{60}{60} = \frac{12}{60}$, and
$\frac{4}{5} = 4 \times \frac{12}{60} = \frac{48}{60}$.

T. Find 6 different expressions for $\frac{3}{5}$.

P. $\frac{3}{5} = \frac{6}{10} = \frac{9}{15} = \frac{12}{20} = \frac{15}{25} = \frac{18}{30} = \frac{21}{35}$.

A similar mode of proceeding is to be followed as to the reduction of 6ths, 7ths, 8ths, &c., to other expressions. The teacher will have observed that the use of the line has been abandoned for a sort of systematic reasoning; yet, whenever the idea is not quite clear in the pupil's mind, it is recommended to have recourse to the senses, by requiring a line to be divided, according as the question may require.

Answers to the Exercises.

Halves.

Ans. 1. $\frac{1}{2} = \frac{6}{12} = \frac{10}{20} = \frac{32}{64} = \frac{43}{86} = \frac{86}{172}$.

2. $\frac{3}{2} = \frac{27}{18} = \frac{57}{38} = \frac{105}{70} = \frac{141}{94} = \frac{357}{238}$.

Thirds.

3. $\frac{1}{3} = \frac{5}{15} = \frac{13}{39} = \frac{19}{57} = \frac{29}{87} = \frac{49}{147}$.

4. $\frac{2}{3} = \frac{36}{54} = \frac{52}{78} = \frac{116}{174} = \frac{172}{258}$.

Fourths.

5. $\frac{1}{4} = \frac{16}{64} = \frac{18}{72} = \frac{24}{96} = \frac{31}{124}$.

6. $\frac{3}{4} = \frac{51}{68} = \frac{75}{100} = \frac{186}{248} = \frac{225}{300}$.

Fifths.

7. $\frac{1}{5} = \frac{11}{55} = \frac{19}{95} = \frac{27}{135} = \frac{60}{300}$.

8. $\frac{3}{5} = \frac{54}{90} = \frac{291}{485} = \frac{402}{670}$.

$\frac{4}{5} = \frac{72}{90} = \frac{388}{485} = \frac{536}{670}$.

Sixths.

9. $\frac{1}{6} = \frac{13}{78} = \frac{19}{114} = \frac{47}{282}$.

$\frac{5}{6} = \frac{65}{78} = \frac{95}{114} = \frac{235}{282}$.

Sevenths.

10. $\frac{1}{7} = \frac{13}{91} = \frac{18}{126} = \frac{24}{168} = \frac{39}{273}$.

$\frac{2}{7} = \frac{26}{91} = \frac{36}{126} = \frac{48}{168} = \frac{78}{273}$.

$\frac{3}{7} = \frac{39}{91} = \frac{54}{126} = \frac{72}{168} = \frac{117}{273}.$

$\frac{4}{7} = \frac{52}{91} = \frac{72}{126} = \frac{96}{168} = \frac{156}{273}.$

$\frac{5}{7} = \frac{65}{91} = \frac{90}{126} = \frac{120}{168} = \frac{195}{273}.$

$\frac{6}{7} = \frac{78}{91} = \frac{108}{128} = \frac{144}{168} = \frac{234}{273}.$

Eighths.

Ans. 11. $\frac{1}{8} = \frac{12}{96} = \frac{15}{120} = \frac{17}{136} = \frac{42}{336}.$

$\frac{3}{8} = \frac{36}{96} = \frac{45}{120} = \frac{51}{136} = \frac{126}{336}.$

$\frac{5}{8} = \frac{60}{96} = \frac{75}{120} = \frac{85}{136} = \frac{210}{336}.$

$\frac{7}{8} = \frac{84}{96} = \frac{105}{120} = \frac{119}{136} = \frac{294}{336}.$

Ninths.

12. $\frac{1}{9} = \frac{13}{117} = \frac{18}{162} = \frac{34}{306}.$

$\frac{2}{9} = \frac{26}{117} = \frac{36}{162} = \frac{68}{306}.$

$\frac{4}{9} = \frac{52}{117} = \frac{72}{162} = \frac{136}{306}.$

$\frac{5}{9} = \frac{65}{117} = \frac{90}{162} = \frac{170}{306}.$

$\frac{7}{9} = \frac{91}{117} = \frac{126}{162} = \frac{238}{306}.$

$\frac{8}{9} = \frac{104}{117} = \frac{144}{162} = \frac{272}{306}.$

Miscellaneous.

13. $\frac{7}{10} = \frac{210}{300}.$ $\frac{14}{15} = \frac{126}{135}.$

$\frac{3}{11} = \frac{33}{121}.$ $\frac{15}{16} = \frac{45}{48}.$

$\frac{11}{12} = \frac{88}{96}.$ $\frac{11}{17} = \frac{22}{34}.$

$\frac{5}{13} = \frac{15}{39}.$ $\frac{17}{18} = \frac{68}{72}.$

$\frac{13}{14} = \frac{91}{98}.$ $\frac{18}{19} = \frac{54}{57}.$

Lesson III.

Reduction of Fractions to the same Denominator.

Teacher. Which are the several expressions for halves?

Pupils. 4ths, 6ths, 8ths, 10ths, 12ths, &c.

T. And which are the several expressions for thirds?

P. 6ths, 9ths, 12ths, &c.

T. If, then, it be required to express $\frac{1}{2}$ and $\frac{1}{3}$ both by other fractions of the same denomination, which will you choose?

P. 6ths, or 12ths, or 18ths, or 24ths, &c.

T. Then reduce $\frac{1}{2}$ and $\frac{1}{3}$ to fractions having the same denominator.

P. $\frac{1}{2} = \frac{3}{6}$, and $\frac{1}{3} = \frac{2}{6}$; or
$\frac{1}{2} = \frac{6}{12}$, and $\frac{1}{3} = \frac{4}{12}$; or
$\frac{1}{2} = \frac{9}{18}$, and $\frac{1}{3} = \frac{6}{18}$; or, &c.

T. How will you proceed in order to reduce $\frac{1}{2}$ and $\frac{1}{4}$ to other fractions having the same denominator?

P. First find the several expressions for $\frac{1}{2}$ and $\frac{1}{4}$, and then choose those which are the same.

T. Reduce $\frac{1}{2}$ and $\frac{1}{4}$ to fractions having the same denominator, or, as it is usual to say, to a *common* denominator.

P. $\frac{1}{2} = \frac{2}{4} = \frac{3}{6} = \frac{4}{8} = \frac{5}{10} = \frac{6}{2}$, &c.

$\frac{1}{4} = \frac{2}{8} = \frac{3}{12} = \frac{4}{16}$, &c.

Hence, the same denominations for $\frac{1}{2}$ and $\frac{1}{4}$ are,—4ths, 8ths, 12ths, 16ths, &c.

Or, $\frac{1}{2} = \frac{2}{4} = \frac{4}{8} = \frac{6}{12} = \frac{8}{16}$, &c.

and $\frac{1}{4} = \frac{2}{8} = \frac{3}{12} = \frac{4}{16}$, &c.

T. What are the numbers 4, 8, 12, 16, &c., of the denominators 2 and 4?

P. They are common multiples of 2 and 4.

T. Hence, if two fractions of different denominators are to be reduced to a common denominator, what must be done?

P. We must find the common multiples of the different denominators,* and reduce the fractions to these denominations.

T. Reduce $\frac{1}{2}$ and $\frac{2}{3}$ to a common denominator.

P. The common multiples of 2 and 3 are,—6, 12, 18, 24, &c.

Hence $\frac{1}{2} = \frac{3}{6} = \frac{6}{12} = \frac{9}{18} = \frac{12}{24}$, &c.

and $\frac{2}{3} = \frac{4}{6} = \frac{8}{12} = \frac{12}{18} = \frac{16}{24}$, &c.

T. Reduce $\frac{1}{2}$ and $\frac{1}{5}$ to a common denominator.

P. The common multiples of 2 and 5 are,—10, 20, 30, 40, &c.

Hence $\frac{1}{2} = \frac{5}{10} = \frac{10}{20} = \frac{15}{30} = \frac{20}{40}$, &c.

and $\frac{1}{5} = \frac{2}{10} = \frac{4}{20} = \frac{6}{30} = \frac{8}{40}$, &c.

---

* See Chap. V, Lesson II.

T. Reduce $\frac{2}{3}$ and $\frac{3}{4}$ to a common denominator.

P. The common multiples of 3 and 4 are,—12, 24, 36, 48, 60, &c.

$$\text{Hence } \tfrac{2}{3} = \tfrac{8}{12} = \tfrac{16}{24} = \tfrac{24}{36} = \tfrac{32}{48} = \tfrac{40}{60}, \text{ \&c.}$$
$$\text{and } \tfrac{3}{4} = \tfrac{9}{12} = \tfrac{18}{24} = \tfrac{27}{36} = \tfrac{36}{48} = \tfrac{45}{60}, \text{ \&c.}$$

T. Which of the several common denominators is the least?

P. 12.

T. Hence, if 2 fractions are to be reduced to the *least* common denominator, what must be done?

P. Find the *least* common multiple of their denominators, and reduce the fractions to that denomination.

T. Reduce $\frac{3}{8}$ and $\frac{4}{5}$ to the least common denominator.

P. The least common multiple of 8 and 5 is 40;

$$\text{hence } \tfrac{3}{8} = \tfrac{15}{40}, \text{ and } \tfrac{4}{5} = \tfrac{32}{40}.$$

T. Reduce $\frac{1}{2}$, $\frac{1}{3}$, and $\frac{1}{4}$ to the least common denominator.

P. The least common multiple of 2, 3, and 4 is 12;

$$\text{hence } \tfrac{1}{2} = \tfrac{6}{12};\ \tfrac{1}{3} = \tfrac{4}{12};\text{ and } \tfrac{1}{4} = \tfrac{3}{12}.$$

Pupils and teacher give questions to the class, after which follow the exercises in Part II.

Answers to the Exercises.

Ans. 1. $\frac{3}{4} = \frac{27}{36}$; $\frac{5}{9} = \frac{20}{36}$.

2. $\frac{4}{5} = \frac{32}{40}$; $\frac{7}{8} = \frac{35}{40}$.

3. $\frac{5}{6} = \frac{35}{42}$; $\frac{6}{7} = \frac{36}{42}$.

4. $\frac{6}{7} = \frac{48}{56}$; $\frac{5}{8} = \frac{35}{56}$.

5. $\frac{8}{9} = \frac{80}{90}$; $\frac{7}{10} = \frac{63}{90}$.

6. $\frac{9}{10} = \frac{54}{60}$; $\frac{7}{12} = \frac{35}{60}$.

7. $\frac{7}{9} = \frac{28}{36}$; $\frac{7}{12} = \frac{21}{36}$.

8. $\frac{8}{9} = \frac{16}{18}$; $\frac{5}{6} = \frac{15}{18}$.

9. $\frac{3}{8} = \frac{9}{24}$; $\frac{5}{6} = \frac{20}{24}$.

10. $\frac{11}{12} = \frac{55}{60}$; $\frac{14}{15} = \frac{56}{60}$.

13. $\frac{30}{60}, \frac{40}{60}, \frac{45}{60}, \frac{48}{60}, \frac{50}{60}, \frac{42}{60}, \frac{55}{60}, \frac{52}{60}, \frac{38}{60}$.

14. $\frac{16}{24}, \frac{20}{24}, \frac{21}{24}, \frac{18}{24}, \frac{12}{24}, \frac{14}{24}$.

15. $\frac{280}{420}, \frac{315}{420}, \frac{336}{420}, \frac{360}{420}$.

16. $\frac{60}{120}, \frac{90}{120}, \frac{100}{120}, \frac{105}{120}, \frac{108}{120}$.

17. $\frac{210}{315}, \frac{270}{315}, \frac{280}{315}, \frac{252}{315}$.

18. $\frac{135}{180}, \frac{144}{180}, \frac{150}{180}, \frac{160}{180}$.

§ 3. ADDITION OF FRACTIONS.

LESSON I. *Addition of Fractions which have the same Denominator.*

Teacher. Tell me what we have been learning concerning fractions.

Pupils. We have learnt,

1st. What halves, thirds, fourths, &c., are;

2nd. To reduce units to fractions;

3rd. To reduce fractions to units;

4th. To reduce a fraction to other denominations;

5th. To reduce several fractions to the least common denominator.

It is of great use to put, from time to time, questions similar to the above to the pupils, and require them to illustrate each case by examples.

Teacher. All that now remains to be learned respecting fractions is, how to add them; to subtract one from another; to multiply and to divide them. We will begin with addition of fractions. Give a question which you think very easy to answer.

Pupils. Add $\frac{1}{2}$ and $\frac{1}{2}$.

T. How much is $\frac{1}{2}+\frac{1}{2}$?

P. $\frac{2}{2}$, or 1.

T. Give some other easy question.

P. Add $\frac{1}{2}$ and $\frac{2}{2}$.

T. How much is $\frac{1}{2}+\frac{2}{2}$?

P. $\frac{3}{2}$, or $1\frac{1}{2}$.

T. Add $\frac{5}{2}$ and $\frac{9}{2}$.

P. $\frac{5}{2}+\frac{9}{2}=\frac{14}{2}=7$.

T. Add $\frac{1}{3}$ and $\frac{1}{3}$.

P. $\frac{2}{3}$.

T. Add $\frac{2}{3}$ and $\frac{2}{3}$.

P. $\frac{4}{3}$ or $1\frac{1}{3}$.

T. Add $\frac{3}{4}$ and $\frac{5}{4}$.

P. $\frac{3}{4} + \frac{5}{4} = \frac{8}{4} = 2$.

T. What kind of fractions have we been adding?

P. Fractions having the same denominator.

T. Add $\frac{3}{5}$ and $\frac{4}{5}$.

P. $\frac{7}{5} = 1\frac{2}{5}$.

Similar questions are to be given in the addition of 6ths, 7ths, 8ths, &c.

Lesson II. *Addition of Fractions having different Denominators.*

Teacher. If now we have to add fractions of different denominators, suppose $\frac{1}{2} + \frac{1}{3}$, what will you do?

Pupils. First reduce $\frac{1}{2}$ and $\frac{1}{3}$ to a common denominator, and then add them; thus:—

$$\frac{1}{2} + \frac{1}{3} = \frac{3}{6} + \frac{2}{6} = \frac{5}{6}.$$

Teacher. Add $\frac{1}{2}$ and $\frac{2}{3}$.

P. $\frac{1}{2} + \frac{2}{3} = \frac{3}{6} + \frac{4}{6} = \frac{7}{6} = 1\frac{1}{6}$.

T. Add $\frac{1}{2}$ and $\frac{1}{4}$.

P. $\frac{1}{2} + \frac{1}{4} = \frac{2}{4} + \frac{1}{4} = \frac{3}{4}$.

T. Add $\frac{1}{2}$ and $\frac{3}{4}$.

P. $\frac{1}{2} + \frac{3}{4} = \frac{2}{4} + \frac{3}{4} = \frac{5}{4} = 1\frac{1}{4}$.

T. Add $\frac{1}{2}$ and $\frac{4}{5}$.

P. $\frac{1}{2} + \frac{4}{5} = \frac{5}{10} + \frac{8}{10} = \frac{13}{10} = 1\frac{3}{10}$.

T. Add $\frac{1}{2}$ and $\frac{5}{6}$.

P. $\frac{1}{2}+\frac{5}{6}=\frac{3}{6}+\frac{5}{6}=\frac{8}{6}=1\frac{2}{6}$.

T. There is something to be remarked here. The answer is $1\frac{2}{6}$, do you know of another expression for $\frac{2}{6}$?

P. Yes $\frac{4}{12}$, or $\frac{6}{18}$, or $\frac{8}{24}$, &c.

T. True; but do you know of another expression for $\frac{2}{6}$ below 12ths?

P. Yes, $\frac{1}{3}$.

T. Now, it is usual to express a fraction by the least expression possible; or, as it is called, to reduce it to *its lowest terms.* Hence $\frac{2}{6}$ reduced to its lowest terms is——

P. $\frac{1}{3}$.

T. Add $\frac{1}{2}$ and $\frac{2}{7}$.

P. $\frac{1}{2}+\frac{2}{7}=\frac{7}{14}+\frac{4}{14}=\frac{11}{14}$.

T. Can $\frac{11}{14}$ be reduced to lower terms?

P. No.

T. Add $\frac{3}{4}$ and $\frac{1}{12}$.

P. $\frac{3}{4}+\frac{1}{12}=\frac{9}{12}+\frac{1}{12}=\frac{10}{12}$.

T. Can $\frac{10}{12}$ be reduced?

P. Yes, $\frac{10}{12}=\frac{5}{6}$.

T. In our next lesson we shall learn what fractions may be reduced to lower terms, and what may not.

Teacher and pupils, as usual, give similar questions to the class, after which the exercises in Part II are taken up.

Answers to the Exercises.

Ans. 1.	$1\frac{5}{12}$.	*Ans.* 13.	$1\frac{33}{54}=1\frac{11}{18}$.
2.	$1\frac{11}{20}$.	14.	$1\frac{23}{27}$.
3.	$1\frac{19}{30}$.	15.	$1\frac{19}{24}$.
4.	$1\frac{29}{42}$.	16.	$1\frac{23}{30}$.
5.	$1\frac{41}{56}$.	17.	$1\frac{17}{24}$.
6.	$1\frac{55}{72}$.	18.	$1\frac{31}{60}$.
7.	$1\frac{71}{90}$.	19.	$1\frac{37}{48}$.
8.	$1\frac{89}{110}$.	20.	$1\frac{22}{30}=1\frac{11}{15}$.
9.	$1\frac{109}{132}$.	21.	$1\frac{43}{80}$.
10.	$\frac{31}{33}$.	22.	$1\frac{76}{90}=1\frac{38}{45}$.
11.	$1\frac{23}{36}$.	23.	$1\frac{48}{70}=1\frac{24}{35}$.
12.	$1\frac{23}{60}$.		

Lesson III.

Teacher. State the different expressions for $\frac{1}{2}$.

Pupils. $\frac{1}{4}=\frac{2}{4}=\frac{3}{6}=\frac{4}{8}=\frac{5}{10}=\frac{6}{12}$, &c.

T. Since $\frac{1}{2}=\frac{6}{12}$, what must be done to bring $\frac{6}{12}$ back again to its lowest denomination $\frac{1}{2}$?

P. Divide its numerator and its denominator by 6.

T. State the different expressions for $\frac{1}{3}$.

P. $\frac{1}{3}=\frac{2}{6}=\frac{3}{9}=\frac{4}{12}=\frac{5}{15}$, &c.

T. Since $\frac{1}{3}=\frac{5}{15}$, what must be done to reduce $\frac{5}{15}$ to its lowest term $\frac{1}{3}$?

P. Divide its numerator and its denominator by 5.

T. Let us now consider the fraction $\frac{14}{18}$, for instance, and see by what number its numerator and denominator must be divided, in order to reduce $\frac{14}{18}$ to its lowest terms.

P. They must be divided by 2, and then $\frac{14}{18}$ will be reduced to $\frac{7}{9}$.

T. Can you reduce $\frac{7}{9}$ to lower terms?

P. No, because there is no number by which both 7 and 9 are divisible.

T. Reduce $\frac{18}{24}$ to its lowest terms.

P. 18 and 24 are both divisible by 2, therefore, $\frac{18}{24}=\frac{9}{12}$; and 9 and 12 are divisible by 3, therefore, $\frac{9}{12}=\frac{3}{4}$; and hence $\frac{18}{24}=\frac{3}{4}$.

T. In general, then, when can a fraction be reduced to lower terms, and when not?

P. It can be reduced when both numerator and denominator are divisible by some one number; and it cannot be reduced when numerator and denominator are not both divisible by the same number.

T. Recollect then to reduce the answers you obtain by adding, subtracting, multiplying, or dividing fractions, always to its lowest term. Can you tell some reason why it is usual to do so?

P. Because in the fraction $\frac{9}{12}$, which, when reduced, is $\frac{3}{4}$; it is easier to imagine the number 1 divided into 4 equal parts, and 3 of these parts taken, than to imagine 1 divided into 12 equal parts, and 9 of these taken.

T. We will now add 3 or more fractions, [writing upon the school-slate.]

Find the sum of $\frac{1}{2}$, $\frac{1}{3}$, and $\frac{1}{4}$.

P. The least common denominator is 12, and $\frac{1}{2}+\frac{1}{3}+\frac{1}{4}=\frac{6}{12}+\frac{4}{12}+\frac{3}{12}=\frac{13}{12}=1\frac{1}{12}$.

T. Add $\frac{1}{2}$, $\frac{2}{3}$, and $\frac{3}{4}$.

P. $\frac{1}{2}+\frac{2}{3}+\frac{3}{4}=\frac{6}{12}+\frac{8}{12}+\frac{9}{12}=\frac{23}{12}=1\frac{11}{12}$.

T. Add $\frac{1}{2}$, $\frac{3}{4}$, and $\frac{4}{5}$.

P. $\frac{1}{2}+\frac{3}{4}+\frac{4}{5}=\frac{10}{20}+\frac{15}{20}+\frac{16}{20}=\frac{41}{20}=2\frac{1}{20}$.

T. Add $\frac{1}{2}$, $\frac{1}{4}$, and $\frac{1}{12}$.

P. $\frac{1}{2}+\frac{1}{4}+\frac{1}{12}=\frac{6}{12}+\frac{3}{12}+\frac{1}{12}=\frac{10}{19}=\frac{5}{6}$.

T. Add $1\frac{1}{2}$, $2\frac{1}{3}$, and $3\frac{1}{4}$.

P. $1\frac{1}{2}+2\frac{1}{3}+3\frac{1}{4}=1\frac{6}{12}+2\frac{4}{12}+3\frac{3}{12}=6\frac{13}{12}=7\frac{1}{12}$.

T. Add $1\frac{4}{5}$ and $1\frac{5}{8}$.

P. $1\frac{4}{5}+1\frac{5}{8}=1\frac{32}{40}+1\frac{25}{40}=2\frac{57}{40}=3\frac{17}{40}$.

Answers to the Exercises.

Ans. 1.	$2\frac{23}{60}$.		*Ans.* 10.	$2\frac{67}{315}$.
2.	$2\frac{13}{60}$.		11.	$3\frac{49}{180}$.
3.	$1\frac{19}{24}$.		12.	$4\frac{2}{15}$.
4.	$2\frac{1}{10}$.		13.	$4\frac{31}{35}$.
5.	$2\frac{14}{27}$.		14.	$10\frac{11}{18}$.
6.	$2\frac{13}{24}$.		15.	$34\frac{1}{8}$.
7.	$2\frac{19}{24}$.		16.	$29\frac{31}{36}$.
8.	$6\frac{7}{15}$.		17.	$60\frac{1}{20}$.
9.	$3\frac{103}{120}$.		18.	$10\frac{2}{15}$.

§ 4. SUBTRACTION OF FRACTIONS.

This lesson is, in all respects, analogous to addition of fractions, and does not involve any new principle;—a few questions relating to the several progressive steps will, therefore, suffice to show the mode of proceeding.

1. *Subtracting a Fraction from Integers.*

Questions. From 1 take $\frac{1}{2}$.
... 2 take $\frac{1}{3}$.
... 3 take $\frac{3}{7}$.
... 17 take $1\frac{3}{5}$.
&c.

2. *Subtracting a Fraction from a Fraction.*

(*a.*) How much is $\frac{1}{2}$ less $\frac{1}{2}$?
... ... $\frac{2}{3}$ ... $\frac{1}{3}$?
... ... $\frac{4}{5}$ ... $\frac{3}{5}$?
... ... $\frac{17}{18}$ ... $\frac{11}{18}$?
... ... $\frac{97}{99}$ ... $\frac{48}{99}$?

(*b.*) What is the difference
between $\frac{1}{2}$ and $\frac{1}{3}$? $\frac{1}{2}-\frac{1}{3}=\frac{3}{6}-\frac{2}{6}=\frac{1}{6}$.
... $\frac{1}{2}$ and $\frac{1}{4}$? $\frac{1}{2}-\frac{1}{4}=\frac{2}{4}-\frac{1}{4}=\frac{1}{4}$.
... $\frac{2}{3}$ and $\frac{3}{4}$? $\frac{2}{3}-\frac{3}{4}=\frac{8}{12}-\frac{9}{12}=\frac{1}{12}$.
... $\frac{4}{5}$ and $\frac{6}{7}$? $\frac{4}{5}-\frac{6}{7}=\frac{28}{35}-\frac{30}{35}=\frac{1}{35}$.
... $\frac{8}{9}$ and $\frac{7}{8}$? $\frac{8}{9}-\frac{7}{8}=\frac{64}{72}-\frac{63}{72}=\frac{1}{72}$.
&c.

3. *Finally, subtracting Integers and Fractions from Integers and Fractions.*

From $2\frac{1}{2}$ take $1\frac{2}{3}$.

Solution. $2\frac{1}{2}-1\frac{2}{3}=\frac{5}{2}-\frac{5}{3}=\frac{15}{6}-\frac{10}{6}=\frac{5}{6}$.

From $3\frac{1}{5}$ take $2\frac{7}{8}$.

Solution. $3\frac{1}{5}-2\frac{7}{8}=\frac{16}{5}-\frac{23}{8}=\frac{128}{40}-\frac{115}{40}=\frac{13}{40}$.

It is to be observed, that *mental* solutions are required, and therefore no questions are to be given involving high numbers, the object being that of unfolding principles, and *preparing* for the rule, and care being necessary not to perplex the minds of the class. Each question may be written on the large school-slate, and the pupils required to find the answers mentally, each lifting up his hand as soon as he has found the answer.

Answers to the Exercises.

1. *Fractions from Integers.*

Ans. 1.	$16\frac{1}{13}$.	*Ans.* 6.	$42\frac{5}{8}$.
2.	$18\frac{27}{72}$.	7.	$42\frac{12}{17}$.
3.	$34\frac{1}{2}$.	8.	$46\frac{19}{47}$.
4.	$7\frac{1}{4}$.	9.	$58\frac{38}{49}$.
5.	$18\frac{3}{7}$.	10.	$66\frac{54}{57}$.

2. *Fractions from Fractions.*

Ans. 1. $\frac{11}{40}$.	*Ans.* 11. $\frac{39}{88}$.
2. $\frac{17}{99}$.	12. $\frac{19}{80}$.
3. $\frac{13}{56}$.	13. $\frac{2}{45}$.
4. $\frac{1}{20}$.	14. $\frac{25}{96}$.
5. $\frac{1}{2}$.	15. $\frac{11}{84}$.
6. $\frac{17}{42}$.	16. $\frac{5}{52}$.
7. 0.	17. $\frac{43}{112}$.
8. $\frac{7}{12}$.	18. 0.
9. $\frac{19}{132}$.	19. $\frac{1}{112}$.
10. $\frac{11}{70}$.	20. $\frac{25}{247}$.

3. *Integers and Fractions from the same.*

Ans. 1. $\frac{7}{8}$.	*Ans.* 6. $2\frac{21}{40}$.
2. $1\frac{12}{35}$.	7. $\frac{5}{6}$.
3. $2\frac{7}{45}$.	8. $1\frac{1}{2}$.
4. $1\frac{19}{28}$.	9. $5\frac{7}{12}$.
5. $3\frac{1}{6}$.	10. $7\frac{7}{18}$.

§ 5. MULTIPLICATION OF FRACTIONS.

The order in which the successive steps of this subject present themselves is:—

1st. To multiply a fraction by integers;
2nd. To multiply integers by fractions;
3rd. To multiply a fraction by fractions;
4th. To multiply integers and fractions by the same.

On the first of these sections no remark is required to be made, since pupils, who have proceeded thus far, find no difficulty in multiplying $\frac{2}{3}$, for instance, by 2, 3, 4, &c., since they have only to take $\frac{2}{3}$ twice, 3 times, 4 times, &c.

The second section might, apparently, be dispensed with, since the *results* are the same, whether 3 be multiplied by $\frac{4}{5}$, or $\frac{4}{5}$ be multiplied by 3. But this principle, clear as it is in the multiplication of integers, is far from being satisfactory when applied to fractions. A child will not hesitate readily to admit that $\frac{2}{3}$, multiplied by 15, gives $\frac{30}{3}$, or 10; but he will not immediately perceive that 15, multiplied by $\frac{2}{3}$, must be 10. The question may be presented thus:—

1. What is the meaning of multiplying a number by another number?

Ans. Taking a number a certain number of times.

2. What is the result if a number be multiplied by 1?

Ans. The number itself.

3. If, then, a number be multiplied by less than 1, (by a part of 1,) what must be the result?

Ans. Less than the number.

4. If a number be multiplied by $\frac{1}{2}$, what must be the result?

Ans. One-half of the number.

5. If multiplied by $\frac{1}{3}$?

Ans. One-third of the number.

6. If by $\frac{2}{3}$?

Ans. One-third of the number taken twice.

7. Hence, what is the meaning, and what is the result, of 15 multiplied by $\frac{2}{3}$?

Ans. To multiply 15 by $\frac{2}{3}$, signifies to take $\frac{1}{3}$ of 15 twice, which evidently gives 10.

The notion, then, to be clearly formed is this—that any number whatever, multiplied by another *less than* 1, must give a result *less than the number* which is to be multiplied (the multiplicand).

Sections 3 and 4 are merely an extension of the same principle.

Lesson I. *Fractions by Integers.*

Teacher. What is the meaning of multiplication?

Pupils. Taking a number a certain number of times.

T. What is the meaning of multiplying $\frac{1}{2}$ by 1, 2, 3, 4, &c.?

P. Taking $\frac{1}{2}$ once, twice, 3 times, 4 times, &c.

T. How much is $\frac{1}{2}$ multiplied by 17?

P. $\frac{17}{2}$, or $8\frac{1}{2}$.

T. What is the meaning of multiplying $\frac{1}{3}$ by 1, 2, 3, 4, &c.?

P. Taking $\frac{1}{3}$ once, twice, 3 times, 4 times, &c.

T. Multiply $\frac{1}{3}$ by 19.

P. $\frac{1}{3}\times 19=\frac{19}{3}=6\frac{1}{3}$.

T. Multiply $\frac{2}{3}$ by 19.

P. $\frac{2}{3}\times 19=\frac{38}{3}=12\frac{2}{3}$.

T. In general, then, what is the meaning of multiplying $\frac{3}{4}$, $\frac{4}{5}$, $\frac{5}{6}$, &c., by 1, 2, 3, 4, &c.?

P. Taking $\frac{3}{4}$, $\frac{4}{5}$, $\frac{5}{6}$, &c., once, twice, 3 times, 4 times, &c.

T. How much is $\frac{3}{7}$ multiplied by 8?

P. 8 times $\frac{3}{7}$, or $\frac{24}{7}$, which are $3\frac{3}{7}$.

T. How much is $\frac{8}{9}\times 15$?

P. 15 times $\frac{8}{9}$, or $\frac{90}{9}$, which are 10.

Lesson II. *Integers by Fractions.*

Teacher. What is the result (product) if a number be multiplied by 1?

Pupils. The number itself.

T. What is the product if a number be multiplied by a number greater than 1?

P. *More* than the number.

T. What, then, must be the product if a number be multiplied by a number which is less than 1?

P. It must be *less* than the number.

T. Name some numbers which are less than 1?

P. $\frac{1}{2}$, $\frac{1}{3}$, $\frac{2}{3}$, $\frac{3}{4}$, $\frac{4}{5}$, &c.

T. If, then, a number be multiplied by $\frac{1}{2}$, or by $\frac{1}{3}$, $\frac{1}{4}$, &c., the product must in each case be ——

P. *Less* than the number.

T. What is the meaning of multiplying by 1?

P. Taking a number once.

T. Hence, what is the meaning of multiplying a number by $\frac{1}{2}$?

P. Taking it $\frac{1}{2}$ times, or to take $\frac{1}{2}$ of it.

T. What is the meaning of multiplying 1, 2, 3, 4, &c., by $\frac{1}{2}$?

P. Taking $\frac{1}{2}$ of 1, 2, 3, 4, &c.

T. How much, then, is 1, 2, 3, 4, &c., multiplied by $\frac{1}{2}$?

P. $\frac{1}{2}$, $\frac{2}{2}$, $\frac{3}{2}$, $\frac{4}{2}$, &c.

T. How much is 17 multiplied by $\frac{1}{2}$?

P. $\frac{17}{2}$, or $8\frac{1}{2}$.

T. What is the meaning of multiplying by $\frac{2}{2}$?

P. Taking $\frac{1}{2}$ of a number twice.

T. How much is 12 multiplied by $\frac{2}{2}$?

P. $\frac{1}{2}$ of 18 taken twice;

$\frac{1}{2}$ of 18 $=$ 9, which taken twice,
$=$ 18.

T. What have you to remark, if a number is to be multiplied by $\frac{2}{2}$?

P. It is the same as multiplying it by 1, since $\frac{2}{2} = 1$.

T. What is the meaning of multiplying by $\frac{1}{3}$?

P. Taking a number $\frac{1}{3}$ times, or to take $\frac{1}{3}$ of it.

T. And what is the meaning of multiplying by $\frac{2}{3}$?

P. Taking $\frac{1}{3}$ of a number twice.

T. Multiply 17 by $\frac{2}{3}$.

P. $17 \times \frac{2}{3}$ is $\frac{1}{3}$ of 17×2;
$\frac{1}{3}$ of $17 = \frac{17}{3}$, which, taken twice,
$= \frac{34}{3} = 11\frac{1}{3}$.

T. What is the meaning of multiplying by $\frac{1}{4}$, $\frac{1}{5}$, $\frac{1}{6}$, $\frac{1}{7}$, &c.?

P. Taking $\frac{1}{4}$, $\frac{1}{5}$, $\frac{1}{6}$, $\frac{1}{7}$, &c. of a number.

T. And what is the meaning of multiplying by $\frac{3}{4}$?

P. Taking $\frac{1}{4}$ of a number 3 times.

T. Multiply 9 by $\frac{3}{4}$.

P. $9 \times \frac{3}{4}$ is $\frac{1}{4}$ of 9×3;
$\frac{1}{4}$ of $9 = \frac{9}{4}$, which $\times 3 = \frac{27}{4} = 6\frac{3}{4}$.

T. What is the meaning of multiplying by $\frac{6}{7}$?

P. Taking $\frac{1}{7}$ of a number 6 times.

T. Are you able to multiply a whole number by a fraction?

P. Yes; we have learnt it in our first lessons on fractions.

T. What kind of questions were those?

P. To take $\frac{2}{3}$, $\frac{3}{4}$, $\frac{4}{5}$, &c., of a number; and this is the same as to multiply a number by $\frac{2}{3}$, or $\frac{3}{4}$, $\frac{4}{5}$, &c.

LESSON III. *Fractions by Fractions.*

Teacher. We must now learn to multiply a fraction by a fraction; and we will begin with ascertaining what is the meaning of multiplying $\frac{1}{2}$, for instance, by $\frac{1}{2}$. You know the meaning and the result if $\frac{1}{2}$ be multiplied by 1.

Pupils. Yes, it means to take $\frac{1}{2}$ once, which gives $\frac{1}{2}$.

T. What, then, is the meaning of multiplying $\frac{1}{2}$ by $\frac{1}{2}$?

P. Taking $\frac{1}{2}$ of $\frac{1}{2}$.

T. How much is that?

One or two of the pupils, perhaps, will answer this question correctly, the majority not. Recourse must then be had to ocular demonstration.

T. Draw a straight line; divide it into halves, each half again into halves; now tell me what part of the whole line one of these second halves is.

P. One-fourth of the line.

T. Apply the same reasoning to the number $\frac{1}{2}$, and tell me what $\frac{1}{2}$ of $\frac{1}{2}$ is?

P. $\frac{1}{4}$.

T. Hence how much is $\frac{1}{2} \times \frac{1}{2}$?

P. $\frac{1}{2}$ of $\frac{1}{2}$, or $\frac{1}{4}$.

T. What is the meaning of multiplying $\frac{1}{3}$ by $\frac{1}{2}$?

P. Taking $\frac{1}{2}$ of $\frac{1}{3}$.

T. You may ascertain this by drawing a line; how will you proceed?

P. Divide a line first into thirds, each third then into halves, and see what part $\frac{1}{2}$ of $\frac{1}{3}$ is of the whole line; it is $\frac{1}{6}$ of it.

T. How much then is $\frac{1}{3} \times \frac{1}{2}$?

P. $\frac{1}{2}$ of $\frac{1}{3}$, or $\frac{1}{6}$.

T. And how much is $\frac{2}{3} \times \frac{1}{2}$?

P. $\frac{1}{2}$ of $\frac{2}{3}$, or $\frac{1}{3}$.

T. What is the meaning of multiplying $\frac{1}{4}$, $\frac{1}{5}$, $\frac{1}{6}$, $\frac{1}{7}$, &c., by $\frac{1}{2}$?

P. Taking $\frac{1}{2}$ of $\frac{1}{4}$, of $\frac{1}{5}$, of $\frac{1}{6}$, of $\frac{1}{7}$, &c.

T. Hence if you wish to learn how to multiply a fraction by $\frac{1}{2}$, you must be able to ascertain readily how much $\frac{1}{2}$ of $\frac{1}{4}$, of $\frac{1}{5}$, &c., is. Need you always take a line and actually divide it?

P. No, we can imagine it.

T. Well, then, ascertain either by drawing a line and dividing it, or by supposing it divided, how much $\frac{1}{2}$ is of $\frac{1}{2}$, of $\frac{1}{3}$, of $\frac{2}{3}$, $\frac{1}{4}$, $\frac{3}{4}$, $\frac{1}{5}$, $\frac{2}{3}$, $\frac{3}{5}$, $\frac{4}{5}$, &c.

The pupils must be able to draw up the following results:—

$\frac{1}{2}$ of $\frac{1}{2} = \frac{1}{4}$.	$\frac{1}{2}$ of $\frac{1}{7} = \frac{1}{14}$.
$\frac{1}{3} = \frac{1}{6}$.	$\frac{3}{7} = \frac{3}{14}$.
$\frac{2}{3} = \frac{1}{3}$.	$\frac{5}{8} = \frac{5}{16}$.
$\frac{1}{4} = \frac{1}{8}$.	$\frac{7}{12} = \frac{7}{24}$.
$\frac{2}{4} = \frac{1}{4}$.	$\frac{9}{10} = \frac{9}{20}$.
$\frac{3}{4} = \frac{3}{8}$.	$\frac{24}{31} = \frac{12}{31}$.

T. A little reasoning will save you a great deal of trouble. For instance, how much is $\frac{1}{2}$ of $\frac{1}{5}$?

P. $\frac{1}{10}$.

T. How much, then, is $\frac{1}{2}$ of $\frac{2}{5}$, $\frac{3}{5}$, $\frac{4}{5}$, $\frac{7}{5}$?

P. $2 \times \frac{1}{10}$, $3 \times \frac{1}{10}$, $4 \times \frac{1}{10}$, $7 \times \frac{1}{10}$, or $\frac{2}{10}$, $\frac{3}{10}$, $\frac{4}{10}$, $\frac{7}{10}$.

T. And if you know how much $\frac{1}{2}$ of $\frac{1}{5}$ is, can you tell me how much $\frac{3}{2}$, $\frac{5}{2}$, $\frac{7}{2}$, of $\frac{1}{5}$ is?

P. Yes, for $\frac{1}{2}$ of $\frac{1}{5} = \frac{1}{10}$;

$\frac{3}{2}$ of $\frac{1}{5} = 3 \times \frac{1}{10} = \frac{3}{10}$.

$\frac{5}{2}$ of $\frac{1}{5} = 5 \times \frac{1}{10} = \frac{5}{10} = \frac{1}{2}$.

$\frac{7}{2}$ of $\frac{1}{5} = 7 \times \frac{1}{10} = \frac{7}{10}$.

T. Hence, how much is $\frac{4}{5}$ multiplied by $\frac{3}{2}$?

P. $\frac{1}{5}$ of $\frac{1}{2} = \frac{1}{10}$;

$\frac{1}{5}$ of $\frac{3}{2} = 3 \times \frac{1}{10} = \frac{3}{10}$; and

$\frac{4}{5}$ of $\frac{3}{2} = 4 \times \frac{3}{10} = \frac{12}{10} = \frac{6}{5} = 1\frac{1}{5}$.

T. How much is $\frac{8}{9} \times \frac{5}{2}$?

P. $\frac{1}{2}$ of $\frac{1}{9} = \frac{1}{18}$;

$\frac{5}{2}$ of $\frac{8}{9} = 8 \times \frac{1}{18} = \frac{8}{18} = \frac{4}{9}$; and

$\frac{5}{2}$ of $\frac{8}{9} = 5 \times \frac{4}{9} = \frac{20}{9} = 2\frac{2}{9}$; therefore

$\frac{8}{9} \times \frac{5}{2} = 2\frac{2}{9}$.

A sufficient number of questions relating to the multiplication of fractions by halves, ought to be given before proceeding further; and it must be remarked, that most children will soon discover the rule, viz. to multiply numerator by numerator, and denominator by denominator; but since it is not the object of this treatise to enter upon rules, but merely to prepare for them, the teacher ought frequently to require of his pupils to explain how they have obtained the result.

From the above, the mode of proceeding as to the multiplication by $\frac{1}{3}$, $\frac{1}{4}$, $\frac{1}{5}$, &c., may be anticipated, and a short outline will be sufficient.

The pupils must ascertain that

$\frac{1}{3}$ of $\frac{1}{2} = \frac{1}{6}$
$\frac{1}{3} = \frac{1}{9}$
$\frac{1}{4} = \frac{1}{12}$
$\frac{1}{5} = \frac{1}{15}$
&c.

$\frac{1}{4}$ of $\frac{1}{2} = \frac{1}{8}$
$\frac{1}{3} = \frac{1}{12}$
$\frac{1}{4} = \frac{1}{16}$
$\frac{1}{5} = \frac{1}{20}$
&c.

$\frac{1}{5}$ of $\frac{1}{2} = \frac{1}{10}$
$\frac{1}{3} = \frac{1}{15}$
$\frac{1}{4} = \frac{1}{20}$
$\frac{1}{5} = \frac{1}{25}$
$\frac{1}{6} = \frac{1}{30}$
&c.

$\frac{1}{6}$ of $\frac{1}{2} = \frac{1}{12}$
$\frac{1}{3} = \frac{1}{18}$
$\frac{1}{4} = \frac{1}{24}$
$\frac{1}{5} = \frac{1}{30}$
$\frac{1}{6} = \frac{1}{36}$
&c.

This done, and committed to memory, is all that is necessary.

Teacher. What is the meaning of multiplying by $\frac{1}{3}$?

Pupils. Taking $\frac{1}{3}$ of a number.

T. What is the meaning of multiplying $\frac{1}{2}$ by $\frac{1}{3}$?

P. Taking $\frac{1}{3}$ of $\frac{1}{2}$.

T. How much is that?

P. $\frac{1}{6}$.

T. How much is $\frac{4}{7}\times\frac{1}{3}$?

P. $\frac{1}{3}$ of $\frac{4}{7}$; $\frac{1}{3}$ of $\frac{1}{7} = \frac{1}{21}$; and
$\frac{1}{3}$ of $\frac{4}{7}=4\times\frac{1}{21}=\frac{4}{21}$; therefore
$\frac{4}{7}\times\frac{1}{3}=\frac{4}{21}$

T. How much is $\frac{8}{9}\times\frac{2}{3}$?

P. $\frac{2}{3}$ of $\frac{8}{9}$. Now $\frac{1}{3}$ of $\frac{1}{9} = \frac{1}{27}$;
and $\frac{1}{3}$ of $\frac{8}{9} = 8 \times \frac{1}{27} = \frac{8}{27}$;
therefore $\frac{2}{3}$ of $\frac{8}{9} = 2 \times \frac{8}{27} = \frac{16}{27}$.
Hence $\frac{8}{9} \times \frac{2}{3} = \frac{16}{27}$.

T. How much is $\frac{7}{8} \times \frac{3}{4}$;

P. $\frac{3}{4}$ of $\frac{7}{8}$. Now $\frac{1}{4}$ of $\frac{1}{8} = \frac{1}{32}$;
therefore $\frac{1}{4}$ of $\frac{7}{8} = 7 \times \frac{1}{32} = \frac{7}{32}$;
and $\frac{3}{4}$ of $\frac{7}{8} = 3 \times \frac{7}{32} = \frac{21}{32}$. *Ans.*

Answers to the Exercises.

1. *Fractions by Integers.*

Ans. 1.	3.	*Ans.* 11.	$27\frac{3}{7}$.
2.	5.	12.	$41\frac{5}{8}$.
3.	7.	13.	73.
4.	12.	14.	78.
5.	317.	15.	29.
6.	$4\frac{1}{2}$.	16.	52.
7.	$13\frac{3}{4}$.	17.	91.
8.	$25\frac{1}{5}$.	18.	112.
9.	$40\frac{5}{6}$.	19.	142.
10.	$32\frac{4}{9}$.	20.	311.

2. *Integers by Fractions.*

Ans. 1.	$31\frac{1}{3}$.	*Ans.* 6.	$343\frac{5}{7}$.
2.	$64\frac{4}{5}$.	7.	$78\frac{2}{7}$.
3.	$93\frac{1}{3}$.	8.	$249\frac{3}{5}$.
4.	$88\frac{8}{9}$.	9.	24.
5.	270.	10.	$91\frac{2}{3}$.

3. *Fractions by Fractions.*

Ans. 1.	$\frac{9}{25}$	*Ans.* 9.	$\frac{10}{27}$
2.	$\frac{16}{49}$	10.	$\frac{16}{39}$
3.	$\frac{25}{64}$	11.	$\frac{7}{40}$
4.	$\frac{36}{49}$	12.	$\frac{11}{27}$
5.	$\frac{3}{10}$	13.	$\frac{21}{400}$
6.	$\frac{5}{16}$	14.	$\frac{1}{100}$
7.	$\frac{15}{28}$	15.	$\frac{21}{100}$
8.	$\frac{6}{77}$	16.	$\frac{81}{100}$

4. *Integers and Fractions by the same.*

Teacher. What is the meaning of multiplying $1\frac{1}{2}$ by $\frac{1}{2}$.

Pupils. Taking $\frac{1}{2}$ of $1\frac{1}{2}$.

T. How much is that?

P. $\frac{1}{2}$ of $1 = \frac{1}{2}$, and $\frac{1}{2}$ of $\frac{1}{2} = \frac{1}{4}$;

therefore $\frac{1}{2}$ of $1\frac{1}{2} = \frac{1}{2} + \frac{1}{4} = \frac{2}{4} + \frac{1}{4} = \frac{3}{4}$.

T. A little reflection will assist you in this multiplication.

How much is $1\frac{1}{2}$?

P. $\frac{3}{2}$.

T. Hence $1\frac{1}{2}$ multiplied by $\frac{1}{2}$ is the same as ——

P. $\frac{3}{2} \times \frac{1}{2}$, which is $\frac{3}{4}$.

T. How much is $1\frac{1}{3} \times \frac{1}{2}$?

P. $1\frac{1}{3} \times \frac{1}{2} = \frac{4}{3} \times \frac{1}{2} = \frac{2}{3}$.

T. How much is $1\frac{2}{3} \times \frac{3}{4}$?

P. $1\frac{2}{3} \times \frac{3}{4} = \frac{5}{3} \times \frac{3}{4} = \frac{15}{12} = 1\frac{3}{12} = 1\frac{1}{4}$.

T. How much is $2\frac{3}{4}\times\frac{4}{5}$.

P. $2\frac{3}{4}\times\frac{4}{5}=\frac{11}{4}\times\frac{4}{5}=\frac{44}{20}=2\frac{4}{20}=2\frac{1}{5}$?

T. And how much is $1\frac{1}{2}\times1\frac{1}{2}$.

P. The same as $\frac{3}{2}\times\frac{3}{2}$, which is $\frac{9}{4}$, or $2\frac{1}{4}$.

T. How much is $1\frac{1}{3}\times1\frac{2}{3}$?

P. $1\frac{1}{3}\times1\frac{2}{3}=\frac{4}{3}\times\frac{5}{3}=\frac{20}{9}=2\frac{2}{9}$.

T. What is the sum, the difference, and the product of $\frac{3}{4}$ and $\frac{3}{5}$?

P. $\frac{3}{4}+\frac{3}{5}=\frac{15}{20}+\frac{12}{20}=\frac{27}{20}=1\frac{7}{20}$, *sum.*

$\frac{3}{4}-\frac{3}{5}=\frac{15}{20}-\frac{12}{20}=\frac{3}{20}$, *difference.*

And $\frac{3}{4}\times\frac{3}{5}=\frac{9}{20}$, *product.*

Questions like the above may be written on the school slate, and the pupils be desired to answer them, performing each operation mentally.

Answers to the Exercises.

Ans. 1.	$\frac{9}{10}$.	*Ans.* 7.	2.
2.	$1\frac{1}{3}$.	8.	$2\frac{11}{12}$.
3.	$\frac{13}{18}$.	9.	$3\frac{3}{20}$.
4.	$5\frac{3}{8}$.	10.	$6\frac{29}{30}$.
5.	$7\frac{7}{15}$.	11.	$4\frac{1}{63}$.
6.	$3\frac{21}{25}$.	12.	$5\frac{31}{40}$.

	Sum.	Difference.	Product.
Ans. 13.	$1\frac{17}{56}$.	$\frac{25}{56}$.	$\frac{21}{56}$.
14.	$3\frac{5}{6}$.	$\frac{5}{6}$.	$3\frac{1}{2}$.
15.	$5\frac{11}{12}$.	$1\frac{5}{12}$.	$8\frac{1}{4}$.
16.	$10\frac{7}{10}$.	$3\frac{7}{10}$.	$25\frac{1}{5}$.

	Sum.	Difference.	Product.
Ans. 17.	$3\frac{15}{56}$	$1\frac{1}{56}$	$2\frac{23}{56}$.
18.	$5\frac{12}{35}$	$\frac{33}{35}$	$6\frac{32}{35}$.
19.	$3\frac{44}{45}$	$\frac{19}{45}$	$3\frac{41}{45}$.
20.	$6\frac{25}{132}$	$\frac{77}{132}$	$9\frac{6}{11}$.

§ 6. DIVISION OF FRACTIONS.

The parts of this subject may be presented in the order observed in the four following sections:—

1. To divide fractions by integers;
2. To divide integers by fractions;
3. To divide fractions by fractions;
4. To divide integers and fractions (mixed numbers) by the same.

The object of division, generally, is to ascertain "how often one number is contained in another number;" and the teacher is strongly recommended to abide by this definition in all cases, although a great effort must be made on the part of the pupils to solve the questions belonging to the first of the four sections.

1. *Fractions by Integers.*

Teacher. What is the meaning of division?

Pupils. Ascertaining how often one number is contained in another number.

T. How often is 1 contained in 1?

P. Once.

T. Hence, how much is $1 \div 1$?

P. 1.

T. Now if a number be greater than 1, can it be contained as often?

P. No; less times.

T. How often then is 2 contained in 1?

P. Less than once; only half as often as 1 is contained.

T. How much is that?

P. $\frac{1}{2}$ times.

T. How much then is $1 \div 2$?

P. $\frac{1}{2}$.

T. How often is 3 contained in 1?

P. $\frac{1}{3}$ times.

T. Hence, how much is $1 \div 3$?

P. $\frac{1}{3}$.

T. How often is 4, 5, 6, 7, &c., contained in 1?

P. $\frac{1}{4}$, $\frac{1}{5}$, $\frac{1}{6}$, $\frac{1}{7}$, &c., times.

T. Hence, how much is $1 \div 4$, $1 \div 5$, $1 \div 6$, $1 \div 7$, &c.?

P. $\frac{1}{4}$, $\frac{1}{5}$, $\frac{1}{6}$, $\frac{1}{7}$, &c.

T. Now, since 1 is contained in 1 once, will 1 be contained as often in a number which is less than 1?

P. No; less times.

T. Hence, how often is 1 contained in $\frac{1}{2}$?

P. Only $\frac{1}{2}$ times.

T. How much then is $\frac{1}{2} \div 1$?

P. $\frac{1}{2}$.

T. And how often is 2 contained in $\frac{1}{2}$?

P. Only $\frac{1}{2}$ times of what 1 is contained in $\frac{1}{2}$, that is $\frac{1}{4}$ times.

T. How much then is $\frac{1}{2} \div 2$?

P. $\frac{1}{4}$.

T. And how much is $\frac{1}{2}$ divided by 3, 4, 5, 6, 7, &c.

P. $\frac{1}{6}, \frac{1}{8}, \frac{1}{10}, \frac{1}{12}, \frac{1}{14}$.

T. How often is 1 contained in $\frac{1}{3}$?

P. Only $\frac{1}{3}$ times.

T. How much then is $\frac{1}{3} \div 1$?

P. $\frac{1}{3}$.

T. How often is 2×1, or 2 contained in $\frac{1}{3}$?

P. $\frac{1}{2}$ of $\frac{1}{3}$ times, $\frac{1}{6}$ times.

T. Hence $\frac{1}{3} \div 3$ is?

P. $\frac{1}{9}$.

T. How much is $\frac{1}{3}$ divided by 4, 5, 6, &c.?

P. $\frac{1}{12}, \frac{1}{15}, \frac{1}{18}$, &c.

T. And how much is $\frac{2}{3}$ divided by 1?

P. $\frac{2}{3}$.

T. Why?

P. Because $\frac{1}{3} \div 1 = \frac{1}{3}$,

therefore $\frac{2}{3} \div 1 = 2 \times \frac{1}{3} = \frac{2}{3}$.

T. How much is $\frac{2}{3} \div 2$?

P. $\frac{1}{3}$; because $\frac{2}{3} \div 1 = \frac{2}{3}$, therefore $\frac{2}{3} \div 2 = \frac{1}{2}$ of $\frac{2}{3} = \frac{1}{3}$.

T. How much is $\frac{1}{4} \div 1$?

P. $\frac{1}{4}$.

T. How much is $\frac{3}{4} \div 2$?

P. $\frac{3}{8}$; because $\frac{3}{4} \div 1 = \frac{3}{4}$, therefore $\frac{3}{4} \div 2 = \frac{1}{2}$ of $\frac{3}{4} = \frac{3}{8}$.

The pupils must continue to give similar solutions for each question which the teacher may give.

These questions may be solved in another way.

Dividing by 2 signifies to take $\frac{1}{2}$ of a number.

...	3	...	...	$\frac{1}{3}$	...
...	4	...	...	$\frac{1}{4}$	...

&c., &c.

Hence, $\frac{1}{2} \div 2 = \frac{1}{2}$ of $\frac{1}{2} = \frac{1}{4}$.

$\frac{1}{2} \div 3 = \frac{1}{3}$ of $\frac{1}{2} = \frac{1}{6}$.

&c.

$\frac{1}{3} \div 5 = \frac{1}{5}$ of $\frac{1}{3} = \frac{1}{15}$.

$\frac{7}{8} \div 9 = \frac{1}{9}$ of $\frac{7}{8} = \frac{7}{72}$.

&c.

Still the first mode of considering division would be preferable.

2. *Integers by Fractions.*

Teacher. How often is 1 contained in 1, 2, 3, 4, &c.?

Pupils. Once, twice, 3 times, 4 times, &c.

T. How often then is $\frac{1}{2}$ contained in 1, 2, 3, 4, &c.?

P. $\frac{1}{2}$ is contained in 1, twice;

... ... 2, 4 times;

... ... 3, 6 times;

... ... 4, 8 times;

&c.

T. How much then is $1 \div \frac{1}{2}$, $2 \div \frac{1}{2}$, $3 \div \frac{1}{2}$, $4 \div \frac{1}{2}$?

P. 2, 4, 6, 8.

T. How often is $\frac{1}{3}$ contained in 1, 2, 3, 4, &c.?

P. $\frac{1}{3}$ is contained in 1, 3 times;
... ... 2, 6 times;
... ... 3, 9 times;
... ... 4, 12 times;
&c.

T. How much then is $1 \div \frac{1}{3}$, $2 \div \frac{1}{3}$, $3 \div \frac{1}{3}$, $4 \div \frac{1}{3}$?

P. 3, 6, 9, 12.

T. Since $\frac{1}{3}$ is contained in 1, 3 times, how often must $\frac{2}{3}$ be contained in 1?

P. Only $\frac{1}{2}$ as often as $\frac{1}{3}$ is contained in 1, that is, $\frac{1}{2}$ of 3, or $\frac{3}{2}$ times.

T. Then how much is $1 \div \frac{2}{3}$?

P. $\frac{3}{2}$, or $1\frac{1}{2}$.

T. How much is $2 \div \frac{2}{3}$?

P. 3; because $2 \div \frac{1}{3} = 6$, therefore $2 \div \frac{2}{3} = \frac{1}{2}$ of $6 = 3$.

T. How much is $3 \div \frac{2}{3}$?

P. $\frac{9}{2}$ or $4\frac{1}{2}$; because
$3 \div \frac{1}{3} = 9$; therefore
$3 \div \frac{2}{3} = \frac{1}{2}$ of $9 = 4\frac{1}{2}$.

T. How much is 1, 2, 3, 4, &c., divided by $\frac{1}{4}$?

P. $1 \div \frac{1}{4} = 4$; $2 \div \frac{1}{4} = 8$; $3 \div \frac{1}{4} = 12$; $4 \div \frac{1}{4} = 16$, &c.

T. How much is $1 \div \frac{3}{4}$?

P. $\frac{4}{3}$ or $1\frac{1}{3}$; because $1 \div \frac{1}{4} = 4$, therefore $1 \div \frac{3}{4} = \frac{1}{3}$ of $4 = \frac{4}{3} = 1\frac{1}{3}$.

T. How much is $7 \div \frac{3}{4}$?

P. $\frac{28}{3}$ or $9\frac{1}{3}$; because $7 \div \frac{1}{4} = 28$, therefore $7 \div \frac{3}{4} = \frac{1}{3}$ of $28 = \frac{28}{3} = 9\frac{1}{3}$.

Similar solutions are required for each of the questions which the teacher and pupils may give to the class. The principle upon which these solutions depend can be made very obvious to the senses; for it is clear that the smaller the measure the greater the result obtained by applying it to an object. Suppose a foot measure to be applied to the length of a table 10 times; it is not necessary actually to apply a $\frac{1}{2}$ foot measure, or $\frac{1}{3}$, $\frac{2}{3}$, $\frac{1}{4}$, $\frac{3}{4}$, &c., of a foot measure; for since a foot measures the object 10 times, a $\frac{1}{2}$ foot must evidently measure it 2×10, that is 20 times; $\frac{1}{3}$ of a foot 3×10, that is 30 times; but a $\frac{2}{3}$ foot measure only $\frac{1}{2}$ of 30, that is 15 times; a $\frac{1}{4}$ foot measure 4×10, that is 40 times; but a $\frac{3}{4}$ foot measure only $\frac{1}{3}$ of 40, or $\frac{40}{3}$, or $13\frac{1}{3}$ times. The intelligence of children of 8 or 9 years old readily seizes these truths, and little effort is then required to apply a similar train of reasoning to abstract numbers. Whenever a child seems perplexed by a question of the kind, he will surely arrive at the true result, if accustomed thus to reason.

Suppose the question; divide 3 by $\frac{7}{9}$, the child begins from what he knows to be true, viz.:—

$$1 \div \frac{1}{9} = 9;$$

$$\text{therefore } 3 \div \frac{1}{9} = 3 \times 9 = 27;$$

$$\text{and hence } 3 \div \frac{7}{9} = \frac{1}{7} \text{ of } 27 = \frac{27}{7} = 3\frac{6}{7}.$$

Again, divide 7 by $\frac{8}{9}$.

Solution. Because $1 \div \frac{1}{9} = 9$;

therefore $7 \div \frac{1}{9} = 7 \times 9 = 63$;

and therefore $7 \div \frac{8}{9} = \frac{1}{8}$ of $63 = \frac{63}{8} = 7\frac{7}{8}$.

These solutions the pupils are required to perform mentally; but the written exercises should be solved as shown above.

Answers to the Exercises.

1. *Fractions by Integers.*

Ans.	1.	$\frac{1}{4}$.	*Ans.*	9.	$\frac{3}{25}$.
	2.	$\frac{4}{35}$.		10.	$\frac{29}{120}$.
	3.	$\frac{5}{54}$.		11.	$\frac{1}{4}$.
	4.	$\frac{3}{35}$.		12.	$\frac{10}{81}$.
	5.	$\frac{7}{88}$.		13.	$\frac{2}{49}$.
	6.	$\frac{2}{9}$.		14.	$\frac{5}{101}$.
	7.	$\frac{1}{33}$.		15.	$\frac{127}{1200}$.
	8.	$\frac{19}{100}$.			

2. *Integers by Fractions.*

Ans.	1.	8.	*Ans.*	7.	$25\frac{3}{5}$.
	2.	$8\frac{3}{4}$.		8.	$4\frac{5}{7}$.
	3.	$18\frac{2}{3}$.		9.	$4\frac{4}{11}$.
	4.	$10\frac{4}{5}$.		10.	8.
	5.	$26\frac{2}{3}$.		11.	7.
	6.	$16\frac{1}{2}$.		12.	7.

Ans. 13.	8.	*Ans.* 17.	13.
14.	10.	18.	4.
15.	11.	19.	5.
16.	12.	20.	9.

3. *Fractions by Fractions.*

The solutions of questions of this section are only an extension of the principles laid down in the former section to fractional numbers. For, since it is known that $\frac{1}{2}$ is contained in 1 twice, it necessarily follows that $\frac{1}{2}$ is contained in $\frac{1}{2}$ of 1 only $\frac{1}{2}$ of twice, that is, once;

or, because $1 \div \frac{1}{2} = 2$;
therefore $\frac{1}{2} \div \frac{1}{2} = \frac{1}{2}$ of $2 = 1$.

Again, because it is known that $\frac{1}{3}$ is contained in 1, 3 times; it follows, that $\frac{1}{3}$ is contained in $\frac{1}{2}$ only $\frac{1}{2}$ of 3 times, that is, $\frac{3}{2}$ times; in other words,

because $1 \div \frac{1}{3} = 3$;
therefore $\frac{1}{2} \div \frac{1}{3} = \frac{1}{2}$ of $3 = \frac{3}{2} = 1\frac{1}{2}$.

And so on with other questions of the kind.

Teacher. How often is $\frac{1}{2}$ contained in 1?

Pupils. Twice.

T. Hence, how often must $\frac{1}{2}$ be contained in $\frac{1}{2}$ of 1, or in $\frac{1}{2}$?

P. Only $\frac{1}{2}$ the number of times it is contained in 1, that is once.

T. How much then is $\frac{1}{2} \div \frac{1}{2}$?

P. 1.

T. Again, how often is $\frac{1}{2}$ contained in 1?

P. Twice.

T. Then how often must $\frac{1}{2}$ be contained in the 3rd part of 1, or in $\frac{1}{3}$?

P. Only the 3rd part of the number of times it is contained in 1, that is $\frac{2}{3}$ times.

T. How much then is $\frac{1}{3} \div \frac{1}{2}$.?

P. $\frac{2}{3}$; because $1 \div \frac{1}{2} = 2$;
therefore $\frac{1}{3} \div \frac{1}{2} = \frac{1}{3}$ of $2 = \frac{2}{3}$.

T. Also, because $1 \div \frac{1}{2} = 2$;
therefore $\frac{1}{4} \div \frac{1}{2}$ is——

P. $\frac{1}{4}$ of 2, or $\frac{2}{4}$, or $\frac{1}{2}$.

T. And, because $1 \div \frac{1}{2} = 2$;
therefore $\frac{1}{5} \div \frac{1}{2}$ is——

P. $\frac{1}{5}$ of $2 = \frac{2}{5}$.

T. And, because $1 \div \frac{1}{2} = 2$;
therefore $\frac{1}{6} \div \frac{1}{2}$ is——

P. $\frac{1}{6}$ of $2 = \frac{2}{6} = \frac{1}{3}$.

T. How much is $\frac{1}{8}$, $\frac{1}{9}$, $\frac{1}{10}$, &c., $\div \frac{1}{2}$?

P. Because $1 \div \frac{1}{2} = 2$;
therefore $\frac{1}{8} \div \frac{1}{2} = \frac{1}{8}$ of $2 = \frac{2}{8} = \frac{1}{4}$.

Again,
because $1 \div \frac{1}{2} = 2$;
therefore $\frac{1}{9} \div \frac{1}{2} = \frac{1}{9}$ of $2 = \frac{2}{9}$; and
because $1 \div \frac{1}{2} = 2$;
therefore $\frac{1}{10} \div \frac{1}{2} = \frac{1}{10}$ of $2 = \frac{2}{10} = \frac{1}{5}$.

T. How much is $\frac{2}{3} \div \frac{1}{2}$?

P. Because $1 \div \frac{1}{2} = 2$;
therefore $\frac{1}{3} \div \frac{1}{2} = \frac{1}{3}$ of $2 = \frac{2}{3}$;
and therefore $\frac{2}{3} \div \frac{1}{2} = 2 \times \frac{2}{3} = \frac{4}{3} = 1\frac{1}{3}$.

T. Divide $\frac{3}{4}$, $\frac{4}{5}$, $\frac{5}{6}$, $\frac{6}{7}$, &c., by $\frac{1}{2}$.

P. Because $1 \div \frac{1}{2} = 2$;
therefore $\frac{1}{4} \div \frac{1}{2} = \frac{1}{4}$ of $2 \div \frac{2}{4} = \frac{1}{2}$;
and therefore $\frac{3}{4} \div \frac{1}{2} = 3 \times \frac{1}{2} = \frac{3}{2} = 1\frac{1}{2}$.

Also, because $1 \div \frac{1}{2} = 2$;
therefore $\frac{1}{5} \div \frac{1}{2} = \frac{1}{5}$ of $2 = \frac{2}{5}$;
and $\frac{4}{5} \div \frac{1}{2} = 4 \times \frac{2}{5} = \frac{8}{5} = 1\frac{3}{5}$.

Again,
because $1 \div \frac{1}{2} = 2$;
therefore $\frac{1}{6} \div \frac{1}{2} = \frac{1}{6}$ of $2 = \frac{2}{6} = \frac{1}{3}$;
and $\frac{5}{6} \div \frac{1}{2} = 5 \times \frac{1}{3} = \frac{5}{3} = 1\frac{2}{3}$.

Also, because $1 \div \frac{1}{2} = 2$;
therefore $\frac{1}{7} \div \frac{1}{2} = \frac{1}{7}$ of $2 = \frac{2}{7}$;
and $\frac{6}{7} \div \frac{1}{2} = 6 \times \frac{2}{7} = \frac{12}{7} = 1\frac{5}{7}$.

T. How much is $\frac{1}{2} \div \frac{1}{3}$?

P. Because $1 \div \frac{1}{3} = 3$;
therefore $\frac{1}{2} \div \frac{1}{3} = \frac{1}{2}$ of $3 = \frac{3}{2} = 1\frac{1}{2}$.

T. And how much is $\frac{1}{2} \div \frac{2}{3}$?

P. Because $1 \div \frac{1}{3} = 3$;
therefore $\frac{1}{2} \div \frac{1}{3} = \frac{1}{2}$ of $3 = \frac{3}{2}$;
and therefore $\frac{1}{2} \div \frac{2}{3} = \frac{1}{2}$ of $\frac{3}{2} = \frac{3}{4}$.

Obs. In this solution, the teacher has to bring forcibly before the minds of his pupils the principle, that, "if a number be contained in another number 6 times, twice the former can be contained in the latter only $\frac{1}{2}$ of 6 times, that is, 3 times."

And 3 times the former, only $\frac{1}{3}$ of 6 times, that is, twice.

4 times the former, only $\frac{1}{4}$ of 6 times, that is, $\frac{6}{4}=\frac{3}{2}$ times.

&c.

And hence, because it has been ascertained that

$\frac{1}{2}\div\frac{1}{3}=\frac{3}{2}$; therefore

$\frac{1}{2}\div 2\times\frac{1}{3}=\frac{1}{2}$ of $\frac{3}{2}=\frac{3}{4}$.

Teacher. How much is $\frac{1}{3}\div\frac{2}{3}$?

Pupils. Because $1\div\frac{1}{3}=3$;

therefore $\frac{1}{3}\div\frac{1}{3}=\frac{1}{3}$ of $3=1$;

and therefore $\frac{1}{3}\div\frac{2}{3}=\frac{1}{2}$ of $1=\frac{1}{2}$.

T. How much is $\frac{1}{4}\div\frac{2}{3}$?

P. Because $1\div\frac{1}{3}=3$;

therefore $\frac{1}{4}\div\frac{1}{3}=\frac{1}{4}$ of $3=\frac{3}{4}$;

and $\frac{1}{4}\div\frac{2}{3}=\frac{1}{2}$ of $\frac{3}{4}=\frac{3}{8}$.

T. How much is $\frac{3}{4}\div\frac{2}{3}$?

P. Because $1\div\frac{1}{3}=3$;

therefore $\frac{1}{4}\div\frac{1}{3}=\frac{1}{4}$ of $3=\frac{3}{4}$;

and $\frac{3}{4}\div\frac{1}{3}=3\times\frac{3}{4}=\frac{9}{4}$;

and therefore $\frac{3}{4}\div\frac{2}{3}=\frac{1}{2}$ of $\frac{9}{4}=\frac{9}{8}=1\frac{1}{8}$.

The pupils must be able to give a similar solution for each question.

The principle remains unaltered, whatever the question may be; for let it be required to divide $\frac{3}{7}$ by $\frac{8}{9}$.

Because $1 \div \frac{1}{9} = 9$;

therefore $\frac{1}{7} \div \frac{1}{9} = \frac{1}{7}$ of $9 = \frac{9}{7}$;

and $\frac{3}{7} \div \frac{1}{9} = 3 \times \frac{9}{7} = \frac{27}{7}$;

and therefore $\frac{3}{7} \div \frac{8}{9} = \frac{1}{8}$ of $\frac{27}{7} = \frac{27}{56}$.

Again, to divide $\frac{5}{8}$ by $\frac{3}{5}$.

Because $1 \div \frac{1}{5} = 5$;

therefore $\frac{1}{8} \div \frac{1}{5} = \frac{1}{8}$ of $5 = \frac{5}{8}$;

hence $\frac{5}{8} \div \frac{1}{5} = 5 \times \frac{5}{8} = \frac{25}{8}$;

and therefore $\frac{5}{8} \div \frac{3}{5} = \frac{1}{3}$ of $\frac{25}{8} = \frac{25}{24} = 1\frac{1}{24}$.

Answers to the Exercises.

Ans. 1.	$1\frac{1}{8}$.	*Ans.* 13.	$\frac{60}{77}$.
2.	$1\frac{1}{15}$.	14.	$\frac{24}{91}$.
3.	$1\frac{1}{3}$.	15.	$3\frac{19}{24}$.
4.	$1\frac{2}{7}$.	16.	$\frac{25}{63}$.
5.	$\frac{3}{4}$.	17.	$2\frac{13}{25}$.
6.	$1\frac{1}{63}$.	18.	$\frac{3}{4}$.
7.	$\frac{63}{64}$.	19.	$\frac{3}{4}$.
8.	$1\frac{23}{40}$.	20.	$\frac{3}{4}$.
9.	$\frac{63}{40}$.	21.	$1\frac{1}{3}$.
10.	$\frac{27}{44}$.	22.	$\frac{3}{4}$.
11.	$1\frac{17}{44}$.	23.	$\frac{3}{4}$.
12.	$1\frac{17}{60}$.	24.	$\frac{3}{4}$.

4. *Integers and Fractions by the same.*

Teacher. What are the kinds of questions you have now learnt to perform in division of fractions?

Pupils. 1. To divide fractions by integers;
2. To divide integers by fractions;
3. To divide fractions by fractions.

T. What do you think now remains to be learnt?

P. To divide integers and fractions by integers and fractions.

T. Give a question of the kind, and let us see if you are able to answer it.

P. Divide $1\frac{1}{2}$ by $1\frac{1}{3}$.

T. Which is the greater of these numbers?

P. $1\frac{1}{2}$.

T. Can you tell whether the answer (quotient) is to be less than 1, or more than 1?

P. It must be more than 1, because $1\frac{1}{3}$ is less than $1\frac{1}{2}$; it must, therefore, be contained more than once in $1\frac{1}{2}$.

T. Well, then, how much is $1\frac{1}{2} \div 1\frac{1}{3}$?

P. $1\frac{1}{2} \div 1\frac{1}{3}$ is the same as
$\frac{3}{2} \div \frac{4}{3}$; and because
$1 \div \frac{1}{3} = 3$, therefore
$\frac{1}{2} \div \frac{1}{3} = \frac{1}{2}$ of $3 = \frac{3}{2}$, and
$\frac{3}{2} \div \frac{1}{3} = 3 \times \frac{3}{2} = \frac{9}{2}$; and therefore
$\frac{3}{2} \div \frac{4}{3} = \frac{1}{4}$ of $\frac{9}{2} = \frac{9}{8} = 1\frac{1}{8}$.

Hence $1\frac{1}{2} \div 1\frac{1}{3} = 1\frac{1}{8}$.

T. Divide $1\frac{1}{3}$ by $1\frac{1}{2}$. Is the quotient to be more or less than 1?

P. Less than 1, because $1\frac{1}{2}$ is more than $1\frac{1}{3}$, and cannot, therefore, be contained once in $1\frac{1}{3}$.

$$\text{Now } 1\frac{1}{3} \div 1\frac{1}{2} = \frac{4}{3} \div \frac{3}{2},$$
$$\text{which is } \frac{8}{9}.$$

T. Hence you can always tell beforehand, whether the quotient is to be less or more than 1; and I advise you to ascertain that before you begin finding the answer, as it will show you whether your answer approaches the truth or not.—Divide $3\frac{3}{4}$ by $4\frac{1}{5}$.

P. The quotient must be less than 1.

$$3\frac{3}{4} \div 4\frac{1}{5} = \frac{15}{4} \div \frac{21}{5}.$$
$$\text{Now } 1 \div \frac{1}{5} = 5;$$
$$\frac{1}{4} \div \frac{1}{5} = \frac{5}{4};$$
$$\frac{15}{4} \div \frac{1}{5} = \frac{75}{4};$$
$$\text{therefore } \frac{15}{4} \div \frac{21}{5} = \frac{75}{84} = \frac{25}{28}. \quad \textit{Ans.}$$

The pupils must give a similar account for each of the questions given to them; and in the exercises of Part II, write out, as shown above, the process by which they have obtained these results.

Answers to the Exercises.

Ans. 1.	$\frac{3}{5}$.	7.	$\frac{45}{91}$.
2.	$1\frac{2}{3}$.	8.	$\frac{9}{14}$.
3.	$1\frac{1}{20}$.	9.	$5\frac{4}{9}$.
4.	$\frac{20}{21}$.	10.	$7\frac{1}{2}$.
5.	$\frac{11}{16}$.	11.	$4\frac{2}{9}$.
6.	$1\frac{1}{3}$.	12.	$\frac{9}{38}$.

Answers to the Promiscuous Questions.

Ans. 1.	$\frac{5}{8}$.	*Ans.* 6.	17.
2.	$3\frac{11}{20}$.	7.	$4\frac{1}{8}$.
3.	$1\frac{2}{7}$.	8.	$4\frac{23}{28}$.
4.	$\frac{7}{108}$.	9.	$\frac{27}{35}$.
5.	$\frac{214}{315}$.	10.	$\frac{19}{360}$.

CHAPTER VII.

PROPORTIONS AND PROGRESSIONS.

§ 1. ARITHMETICAL PROPORTION, OR EQUI-DIFFERENCE.

Teacher. What is the difference between two equal numbers?

Pupils. Nothing.

T. Name two numbers whose difference is 1.

P. 1 and 2.

T. Whose difference is 2——

P. 1 and 3.

T. Whose difference is 3——

P. 1 and 4.

T. Which of these is the greater; the first or the second?

P. The second, namely 4.

T. Name 2 other numbers whose difference is also 3; the second being the greater of the two.

P. 5 and 8.

T. Hence, of the four numbers, 1, 4, 5, 8, it may be said that the difference between the 1st and 2nd is the same as——

P. The difference between the 3rd and 4th.

T. I will write this upon the slate. [Writing.]

$$1 \sim 4 = 5 \sim 8.$$

Besides this, there is something else to be remarked concerning these 4 numbers. What is it?

P. That the second is by as much greater than the 1st, as the 4th is greater than the 3rd.

T. Can these 4 numbers be so placed, that the 1st is by as much greater than the 2nd, as the 3rd is greater than the 4th.

P. Yes, thus: $4 \sim 1 = 8 \sim 5$.

T. Find 4 other numbers of which the same may be said as of these 4.

P. $2 \sim 5 = 7 \sim 10$, or

$5 \sim 2 = 10 \sim 7$.

T. What is the difference between the 1st and 2nd pair of these numbers?

P. 3.

T. Hence it may be said that their difference, 3, is *common* to each pair, or that they have a *common* difference.

Find two pairs of numbers whose common difference is 4.

P. 6, 10, and 11, 15.

T. How am I to place these 4 numbers, so that the order before mentioned may be observed?

P. $6 \sim 10 = 11 \sim 15$, or
$10 \sim 6 = 15 \sim 11$.

T. Can they be placed so that the difference between each pair shall still be 4, and this order not be observed?

P. Yes; $6 \sim 10 = 15 \sim 11$, or
$10 \sim 6 = 11 \sim 15$.

T. Now the 1st of these is less than the 2nd, but the 3rd is ——

P. Greater than the 4th.

T. Or, the 1st is greater than the 2nd, but the 3rd is ——

P. Less than the 4th.

T. Now I wish you to find 2 pair of numbers whose common difference is 5, and to place them so as that the order before mentioned may be observed.

P. $5 \sim 10 = 11 \sim 16$, or
$10 \sim 5 = 16 \sim 11$.

T. Find 2 pair whose common difference is 6.

P. $7 \sim 13 = 15 \sim 21$, or
$13 \sim 7 = 21 \sim 15$.

T. Two pair of numbers, similarly related, and placed in such an order, are said to form an *arithmetical proportion*, or *equi-difference.*

Now state how 2 pair of numbers must be related, and how they must be placed, in order to form an arithmetical proportion.

P. The difference between each pair must be

the same; and they must be so placed that according as the 1st is greater or less than the 2nd, the third is likewise greater or less than the 4th.

T. Find 6 arithmetical proportions.

P. $7 \sim 3 = 9 \sim 5$;

$13 \sim 17 = 25 \sim 29$;

$18 \sim 12 = 20 \sim 26$;

$33 \sim 54 = 17 \sim 38$;

$49 \sim 15 = 86 \sim 52$;

$100 \sim 80 = 30 \sim 10$.

T. Do $17 \sim 15 = 39 \sim 35$ form an arithmetical proportion?

P. No; because they have not a common difference.

T. Do $18 \sim 5 = 10 \sim 23$ form an arithmetical proportion?

P. No; because though they have a common difference, yet the 1st is greater than the 2nd; but the 3rd is less than the 4th, which must not be.

T. If, now, the first 3 numbers, or *terms*, as they are usually called, be known, you will be able, I think, to find the 4th term. For instance, the difference between 3 and 7 is the same as the difference between 9 and what other number?

P. 5 or 13.

T. True; but recollect that the four numbers must form an arithmetical proportion, and that I mentioned the lesser first.

P. It must be 13, and not 5.

T. Hence $3 \sim 7 = 9 \sim 13$;
but had I said the difference between

7 and 3 = the difference between 9

and what other number? your answer would have been ——

P. 5.

T. Let us try another question. [Writing on the slate.]

$$13 \sim 9 = 19 \sim$$

Is the 4th term to be less or greater than 19?

P. Less than 19, because the 1st term, 13, is greater than the 2nd term, 9; hence the 3rd term, 19, must be greater than the 4th term, or the 4th term must be less than 19.

T. And by how much must it be less than 19?

P. By as much as 9 is less than 13; that is, by 4.

T. How much, then, is the 4th term?

P. $19 \sim 4$, or 15.

T. Hence $13 \sim 9 = 19 \sim 15$.

What, then, must be inquired into, in order to find the 4th term?

P. First, whether it is to be greater or less than the 3rd; and then, by how much it is to be greater or less.

T. Find a fourth number which shall form an equi-difference with 5, 11, 17.

P. The difference between 5 and $11 = 6$; and since 5 is less than 11, the 3rd term, 17, must be less than the 4th; that is, it must be less than the 4th by 6; the number, therefore, is $17+6=23$.

$$\text{Hence } 5 \sim 11 = 17 \sim 23.$$

T. Find a 4th number which shall form an equi-difference with the numbers $\frac{1}{2}$, $\frac{1}{3}$, $\frac{1}{4}$.

P. The difference between

$$\tfrac{1}{2} \text{ and } \tfrac{1}{3} = \tfrac{3}{6} \sim \tfrac{2}{6} = \tfrac{1}{6};$$

and since $\frac{1}{2}$ is greater than $\frac{1}{3}$ by $\frac{1}{6}$, therefore $\frac{1}{4}$ must be greater than the 4th term by $\frac{1}{6}$.

$$\text{Hence } \tfrac{1}{4} - \tfrac{1}{6} = \tfrac{3}{12} - \tfrac{2}{12} = \tfrac{1}{12}, \text{ the 4th term;}$$
$$\text{and therefore } \tfrac{1}{2} \sim \tfrac{1}{3} = \tfrac{1}{4} \sim \tfrac{1}{12}.$$

A similar solution is required for each question; and since the mind has to dwell some time upon each question, the teacher is advised to write the 3 numbers each time upon the school-slate, or let the pupils write the question upon their own slates.

Answers to the Exercises.

Ans. 4.	74.	*Ans.* 10.	48.
5.	100.	11.	$\frac{7}{12}$.
6.	33.	12.	$\frac{35}{36}$.
7.	51.	13.	$\frac{11}{40}$.
8.	61.	14.	$4\frac{1}{12}$.
9.	29.	15.	$80\frac{5}{12}$.

§ 2. ARITHMETICAL PROGRESSION.

Teacher. Name 4 numbers whose successive differences are equal.

Pupils. 1, 2, 3, 4.

T. What is the common difference?

P. 1.

T. Name 4 other numbers whose successive differences are 2.

P. 1, 3, 5, 7.

T. What difference is there between these four numbers, and 4 numbers forming an equi-difference?

P. The numbers 1, 3, 5, 7, form an equi-difference; but the difference, 2, is not only common to the 1st and 2nd pairs, but also is the difference between the 2nd and 3rd terms.

T. Name 4 other numbers whose successive differences are 3.

P. 1, 4, 7, 10.

T. Can there be more than 4 numbers having this property?

P. Yes, as many as you please; thus:—

1, 4, 7, 10, 13, 16, 19, 22, &c.

T. A series of numbers, whose successive differences are equal, is called an *arithmetical progression.* Form an arithmetical progression, the numbers, or *terms*, as they are called, having a common difference, 4.

P. 1, 5, 9, 13, 17, 21, 25, &c.

T. You made 1 the first term; could the series begin with 2?

P. Yes; 2, 6, 10, 14, 18, 22, &c.

T. With 3?

P. Yes; 3, 7, 11, 15, &c.

T. In short, it might begin with any number. If, then, I tell you to form a progression, beginning with 5, and the difference of whose terms is 7, can you do it?

P. Yes; 5, 12, 19, 26, 33, &c.

T. If the 1st term of a progression be 2, and the 2nd term 7; what is the difference of the terms?

P. 5.

T. And what is the progression?

P. 2, 7, 12, 17, 22, 27, &c.

T. The first term is 9; the 2nd term 16; what is the difference, and what the following terms?

P. The difference is 7, and the terms are,—

9, 16, 23, 30, 37, &c.

T. If the 1st term be 100; the 2nd, 97; what is the difference, and what are the terms of the progression?

P. The difference is 3, and the progression is—

100, 97, 94, 91, 88, 85, 82, &c.

T. What is there to be remarked concerning the

terms of this series, when compared with those of the former?

P. The terms in this series are decreasing; in the former increasing.

T. Hence, an arithmetical progression may be——

P. Either *increasing* or *decreasing*.

T. The 1st term of a series is 85; the 2nd 73. Is the series decreasing or increasing?

P. Decreasing, because the 2nd term is less than the 1st.

T. What are the terms of the series?

P. The difference of the terms is 12; hence the terms are

85, 73, 61, 49, 37, &c.

T. If the 1st term be 1, and the common differ ence be likewise 1, what is the series?

P. 1, 2, 3, 4, 5, &c.

T. What is the 10th term of this series?

P. 10.

T. If the 1st term be 1, the common difference 2, what is the 10th term?

P. 19.

T. How have you found this?

P. We have written 10 terms of this progression, thus:—

1, 3, 5, 7, 9, 11, 13, 15, 17, 19.

T. And, had I asked you to find the 100th term, what would you have done?

P. We should have drawn up 100 terms of the progression.

T. This would have been rather troublesome; now, if you will pay attention and reflect a little, we shall be able to find any term whatever, without first writing down all the previous terms.

[Writing upon the slate:]—

1.	2.	3.	4.	5.	6.	7.
1.	3.	5.	7.	9.	11.	13.
1.	4.	7.	10.	13.	16.	19.
2.	7.	12.	17.	22.	27.	32.

What have I written here upon the slate?

P. 7 terms of each of 4 progressions.

T. The first of these may serve to show us the number of terms in each of the others; we will begin with the 2nd series. Its first term is 1; what is the difference of the terms?

P. 2.

T. Now, tell me how the 2nd term 3 is obtained?

P. By adding the difference, 2, to the 1st term, 1.

T. Is this true for each of the other two progressions?

P. Yes; for, in the third, the 1st term is 1, and the difference 3; 2nd term is $1+3=4$; again, in the last progression, the 1st is 2, difference 5; and the second term is $2+5=7$.

T. Hence the 2nd term of every progression is equal to ——

P. To the 1st term, more the difference.

T. And how is the 3rd term obtained from the 2nd?

P. By adding the common difference 2 to it; thus the 2nd term is 3, difference 2; hence the 3rd term is 3 + 2 = 5.

T. Now, recollect, we ascertained before that the 2nd term is the same as the 1st term, + the difference; hence the 3rd term must be ——

P. The same as the 1st term, more the difference; and more the difference again, that is, more twice the difference.

T. Repeat what you have ascertained concerning the 2nd and 3rd term.

P. The 2nd term = 1st term + the difference;
3rd term = 1st term + twice the diff.

T. See if this be true for each of the other 2 progressions.

P. Yes, it is.

T. Now observe how the 4th term is obtained from the 3rd.

P. By adding the common difference to it.

T. And how is it obtained from the 1st term?

P. By adding 3 times the difference to the 1st term.

T. Hence the 4th term is equal to ——

P. The 4th term is equal to the 1st term + 3 times the common difference.

T. I think you will now be able to find out how the following terms are obtained from the 1st term.

The pupils will thus find, that

The 2nd term = 1st term, + once the common difference.

The 3rd term = 1st term, + twice the common difference.

The 4th term = 1st term, + 3 × the common difference.

The 5th term = 1st term, + 4 × the common difference.

The 6th term = 1st term, + 5 × the common difference.

&c.

The 10th term = 1st term, + 9 × the common difference.

&c., &c.,

if the series be *increasing;* but if decreasing, the difference must be *subtracted* from, instead of *added* to, the 1st term.

T. If the 1st term be 1, and the common difference 3, what will be the 10th term?

P. The 10th term = 1st term, + 9 × the difference; now the 1st term = 1, difference = 3; hence the 10th term = 1 + 9 × 3 = 28.

T. By what means will you ascertain if this be true?

P. By writing 10 terms of the progression 1, 4, 7, &c.

The pupils should be called upon to verify their answers for each question.

Answers to the Exercises.

Ans. 3. 1; 7; 16; 19; 25; 31.

4. 100; 91; 82; 73; 64; 55.

5. $\frac{1}{2}$; $\frac{4}{5}$; $1\frac{1}{10}$; $1\frac{2}{5}$; $1\frac{7}{10}$; 2.

6. 20; $19\frac{1}{3}$; $18\frac{2}{3}$; 18; $17\frac{1}{3}$; $16\frac{2}{3}$.

7. See § 2. *Arithmetical Progression.*

8.	43.	*Ans.* 12.	$27\frac{4}{7}$.
9.	97.	13.	80.
10.	$5\frac{1}{2}$.	14.	293.
11.	6.	15.	$1\frac{2}{5}$.

§ 3. GEOMETRICAL PROPORTION.

Teacher. What is meant by comparing one thing with another?

Pupils. Ascertaining in what respect the 2 things are alike, or in what respect they differ from each other.

T. Hence, if we had to compare two numbers with each other, what would you do?

P. Ascertain whether they are equal to each other, or not.

T. Compare the numbers 2 and 6.

P. They are not equal to each other.

T. And if 2 numbers are not equal to each other, what may be said of them respectively?

P. That the one is greater than the other; or that the one is less than the other.

T. By what operation do you ascertain by how much 6 is greater than 2?

P. By subtracting the lesser number 2 from the greater number 6.

T. And what is the result of this operation called?

P. The difference between 6 and 2, which is 4.

T. Is there *any* other way of comparing these two numbers with each other?

P. Yes, by ascertaining how often the one is contained in or contains the other.

T. By what operation do you ascertain how often 2 is contained in 6, or 6 in 2?

P. By dividing 6 by 2, or 2 by 6.

T. And what is the result of this operation called?

P. The quotient of $6 \div 2$, which is 3, or of $2 \div 6$, which is $\frac{2}{6}$ or $\frac{1}{3}$.

T. Name 2 numbers, of which the difference is the same as that between 2 and 6.

P. 12 and 16.

T. Do you recollect what was said of 4 numbers similarly related as the numbers 2, 6, 12, 16?

P. Yes; they form an equi-difference, or arithmetical proportion.

T. And can you find more pairs of numbers, whose difference is the same as that between 2 and 6?

P. Yes, as many as you please; 14 and 18; 20 and 24, &c.

T. Now, find 2 numbers, of which the quotient is the same as that of 6 divided by 2.

P. 12 divided by 4; or $18 \div 6$; $30 \div 10$, &c.

T. Hence what may be said of the 4 numbers 6, 2, 12, 4?

P. That the 1st 6 divided by the 2nd 2, is the same as the 3rd 12 divided by the 4th 4.

T. In short, that $6 \div 2 = 12 \div 4$. Now find 2 numbers whose quotient is the same as that of $1 \div 2$.

P. $3 \div 6$, or $4 \div 8$, or $5 \div 10$, &c.

T. How did you find these numbers?

P. Since $1 \div 2 = \frac{1}{2}$, any two numbers of which the one is $\frac{1}{2}$ of the other must, when the less is divided by the greater, have a quotient $= \frac{1}{2}$; such numbers are $3 \div 6$, $4 \div 8$, $5 \div 10$.

T. In short, then, $1 \div 2 = 3 \div 6$.

Find two numbers whose quotient is the same as that of $2 \div 1$.

P. Because $2 \div 1 = 2$; any two numbers, of which the one is twice as much as the other, must, when the greater is divided by the less, have a quotient = 2; such are $6 \div 2$, $8 \div 4$, $10 \div 5$, &c.

T. We have found before that

$1 \div 2 = 3 \div 6$; and now,

that $2 \div 1 = 6 \div 3$.—What conclusion can you make?

P. That of 4 numbers, if the 1st divided by the 2nd, is equal to the 3rd divided by the 4th, then also is the 2nd divided by the 1st, equal to the 4th divided by the 3rd.

T. Any 4 numbers having this property are said to be *proportional numbers.*

So that it may be said,

1 has the same proportion to 2,
as 3 has ... ... to 6,
or, as 4 has ... ... to 8.
&c.

And that,

2 has the same proportion to 1,
as 6 has ... ... to 3,
or, as 8 has ... ... to 4.
&c.

And this is shortly written thus:—

1 : 2 : : 3 : 6.
1 : 2 : : 4 : 8.
2 : 1 : : 6 : 3.
And 2 : 1 : : 8 : 4.

Which is read thus:

as 1 is to 2, so is 3 to 6;
as 2 is to 1, so is 6 to 3,
&c.

Now, find 2 numbers which have the same proportion that 1 has to 3.

P. 2 to 6; 3 to 9; 4 to 12, &c.; or any numbers of which the first is $\frac{1}{3}$ of the other.

T. How have you ascertained this?

P. By dividing 1 by 3, which is $\frac{1}{3}$, and any two numbers of which the first is $\frac{1}{3}$ of the other, are in the proportion of 1 to 3.

T. Find two numbers in the proportion of 3 to 1.

P. 6 to 2; 9 to 3; 12 to 4, &c.; or any two numbers of which the first is 3 times as much as the other.

T. Hence, to ascertain two numbers, which have the same proportion to each other as two other numbers, what must be done?

P. Divide the one by the other; and find two numbers, which, when divided, have the same quotient.

T. Speaking of proportional numbers, this quotient is called the *ratio*, which the two numbers have to each other. What is the ratio of 6 to 12?

P. $\frac{6}{12}$ or $\frac{1}{2}$.

T. And what is the ratio of 12 to 6?

P. $\frac{12}{6}$ or 2.

T. And of 4 numbers which form a proportion, according as the 1st is greater or less than the 2nd, the 3rd must be ——

P. Greater or less than the 4th.

T. Hence, as 1 : 4 : : 5 to what other number? —Is it to be less or greater than 5?

P. It must be greater than 5, since the first term 1, is less than 4, the 2nd.

T. What is the number?

P. 20; since the ratio of 1 to 4 is $\frac{1}{4}$, hence the number of which 5 is $\frac{1}{4}$, is the number required; which, therefore, is 4×5, or 20.

T. This is called finding the 4th *proportional* to the 3 numbers 1, 4, 5. Find the 4th proportional to 4, 1, 5.

P. Since $\frac{4}{1} = 4$, the ratio of 5 to the number must be 4, that is, 5 must be 4 times as much as the number, which is, therefore, $\frac{1}{4}$ of 5, or $1\frac{1}{4}$; hence, 4 : 1 : : 5 : $1\frac{1}{4}$.

T. As 2 is to 3, so is 4 to what 4th number?

P. Because 2 is $\frac{2}{3}$ of 3, 4 must be $\frac{2}{3}$ of the number; hence $\frac{1}{3}$ of the number $= \frac{1}{2}$ of 4; and therefore the number must be $\frac{1}{2}$ of 4 taken 3 times, which is 6; therefore,

2 : 3 : : 4 : 6.

N.B. The pupils will not find it difficult to ascertain the 4th proportional, when the 1st term is either a multiple of the 2nd, or the 2nd a multiple of the first. When, however, the ratio of the first two terms is $\frac{2}{3}$, $\frac{3}{4}$, $\frac{2}{5}$, $\frac{3}{5}$, $\frac{4}{5}$, $\frac{5}{6}$, &c., that is, *fractional*, they will naturally experience greater difficulty.

Before entering then upon similar questions, it will be advisable to go through the following exercises.

6 is the half of what number?

Ans. Of 2×6; that is, of 12.

6 is the third part of what number?

Ans. Of 3×6; that is, of 18.

6 is $\frac{2}{3}$ of what number?

Ans. Of the half of 6 taken 3 times; that is, of 9.

6 is $\frac{1}{4}$ of what number?

Ans. Of 4×6; that is, of 24.

6 is $\frac{3}{4}$ of what number?

Ans. Of $\frac{1}{3}$ of 6 taken 4 times; that is, of 8.

6 is $\frac{1}{5}$ of what number?

Ans. Of 5×6; that is, of 30.

6 is $\frac{2}{5}$ of what number?

Ans. Of $\frac{1}{2}$ of 6 taken 5 times; that is, of 15.

6 is $\frac{3}{5}$ of what number?

Ans. Of $\frac{1}{3}$ of 6 taken 5 times; that is, of 10.

6 is $\frac{4}{5}$ of what number?

Ans. Of $\frac{1}{4}$ of 6 taken 5 times; that is

$$\frac{6}{4} \times 5 = \frac{30}{4} = 7\frac{1}{2}.$$

6 is $\frac{5}{7}$ of what number?

Ans. Of $\frac{1}{5}$ of 6 taken 7 times; that is,

$$\frac{6}{5}\times 7=\frac{42}{5}=8\tfrac{2}{5}.$$

11 is $\frac{3}{8}$ of what number?

Ans. Of $\frac{1}{3}$ of 11 taken 8 times; that is,

$$\frac{11}{3}\times 8=\frac{88}{3}=29\tfrac{1}{3}.$$

&c., &c.

These and similar questions having been duly practised, no difficulty will be met with in the following:—

Teacher. As 7 is to 1, so is 13 to——

Pupils. $1\frac{6}{7}$. Because the 1st term is 7 times as much as the 2nd, the 3rd must be 7 times as much as the 4th. Now 13 is 7 times $\frac{13}{7}$, which is $1\frac{6}{7}$.

T. As 7 is to 2, so is 8 to——

P. $2\frac{2}{7}$; because 7 is $\frac{7}{2}$ of 2, therefore 8 must be $\frac{7}{2}$ of the number. Now 8 is $\frac{7}{2}$ of $\frac{1}{7}$ of 8 taken twice; that is,

$$\frac{8}{7}\times 2, \text{ or } \frac{16}{7}=2\tfrac{2}{7}.$$

T. If your answer, $2\frac{2}{7}$, be correct, what must be——

P. 7, divided by 2, must give the same quotient as 8, divided by $2\frac{2}{7}$:—

$$7\div 2=\frac{7}{2}, \text{ and } 8\div 2\tfrac{2}{7}=8\div\frac{16}{7}=\frac{7}{2}.$$

Hence the answer is correct.

Answers to the Exercises.

N.B. The answers to the first 7 questions are contained in § 3. *Geometrical Proportions.*

Ans. 8.

Ratio of 3 to 7 $=\frac{3}{7}$.	Ratio of $\frac{1}{3}$ to $\frac{1}{2}=\frac{2}{3}$.
7 to 3 $=\frac{7}{3}$.	$\frac{2}{5}$ to $\frac{3}{5}=\frac{2}{3}$.
8 to 11 $=\frac{8}{11}$.	$\frac{3}{5}$ to $\frac{2}{5}=\frac{3}{2}$.
11 to 8 $=\frac{11}{8}$.	$\frac{3}{4}$ to $\frac{4}{5}=\frac{15}{16}$.
$\frac{1}{2}$ to $\frac{1}{3}=\frac{3}{2}$.	$\frac{4}{5}$ to $\frac{3}{4}=\frac{16}{15}$.

Ans.	9.	1	:	5	: :	7	:	35.
	10.	5	:	1	: :	9	:	$1\frac{4}{5}$.
	11.	4	:	12	: :	13	:	39.
	12.	12	:	4	: :	13	:	$4\frac{1}{3}$.
	13.	15	:	60	: :	3	:	12.
	14.	9	:	72	: :	14	:	112.
	15.	72	:	9	: :	14	:	$1\frac{3}{4}$.
	16.	17	:	51	: :	23	:	69.
	17.	18	:	90	: :	85	:	425.
	18.	19	:	114	: :	105	:	630.

19. Of 15.
20. Of 16.
21. Of $10\frac{2}{3}$.
22. Of $8\frac{3}{4}$.
23. Of 21.
24. Of $31\frac{1}{4}$.
25. Of 20.

Ans.	26.	2	:	3	: :	9	:	$13\frac{1}{2}$.
	27.	3	:	2	: :	9	:	6.
	28.	3	:	4	: :	15	:	20.
	29.	4	:	3	: :	15	:	$11\frac{1}{4}$.
	30.	4	:	5	: :	18	:	$22\frac{1}{2}$.
	31.	5	:	4	: :	18	:	$14\frac{2}{5}$.
	32.	6	:	7	: :	5	:	$5\frac{5}{6}$.
	33.	$\frac{1}{2}$	:	5	: :	$\frac{1}{3}$	:	$3\frac{1}{3}$.
	34.	$\frac{3}{4}$	:	$\frac{4}{5}$	: :	6	:	$6\frac{2}{5}$.

III.

RITCHIE'S PRINCIPLES OF GEOMETRY.

Familiarly Illustrated, and applied to a variety of useful purposes. Designed for the Instruction of Young Persons. Second Edition, revised and enlarged. 12mo., with 150 Wood-cuts. 3s. 6d. cloth.

THIS book is intended to conduct the student through the elements of plane and solid geometry in a method uniting theory with practice, each illustrating the other. As the fundamental properties of the science are communicated, they are applied, in examples, to the mixed sciences and the mechanical arts. By this means, the two principal advantages of mathematical learning are combined: 1st, The discipline of the reasoning powers by the habit of rigorous demonstration; and, 2nd, The possession and employment of a key to many of the most important arts of life.

To the young engineer, this book is peculiarly adapted, as it not only shows the construction and use of the instruments employed in surveying—such as the diagonal scale, theodolite, and cross-staff—but gives a variety of problems in surveying and levelling, the determination of heights and distances, mensuration of surfaces and solids, and the elements of trigonometry. The nature of geometrical analysis is also explained, and the pupil is introduced to the most important principles of optics and astronomy. The work is interspersed with numerous exercises, carefully adapted to the state of the learner's knowledge at each step of his progress.

The *Athenæum*, in reviewing this book, describes it as "the best introduction to geometry that exists in our language; it is just the work by which a parent may be enabled to instruct his children in the elementary principles of the science, though his own knowledge of it be neither deep nor extensive. The practical applications which are added, must render the study very delightful to the young, since the exercises on the principles will be found as amusing as the ordinary sports of childhood."

IV.

RITCHIE'S PRINCIPLES of the DIFFERENTIAL AND INTEGRAL CALCULUS,

Familiarly Illustrated and applied to a variety of useful purposes. Second Edition, Revised by J. A. SPENCER, B.A., Assistant Mathematical Master in University College School. 12mo., with Diagrams. 4s. 6d. cloth.

THIS work is intended to facilitate the introduction of the young mathematician to the differential and integral calculus, 1st, by requiring no further preparation than the elements of geometry and plane trigonometry, with algebra as far as quadratic equations inclusive; 2nd, by exhibiting to him familiar illustrations of its application to practical purposes. Many students are deterred from the study of the calculus by the formidable array of abstruse analytical formulæ which beset its entrance; and others, who have succeeded in overcoming the first difficulties, relinquish the task from inability to perceive its utility. The author, by commencing with simple arithmetical and geometrical examples of the nature of variation, limits, and functions, initiates the student into an acquaintance with the fundamental principles; and, by treating on the processes of integration concurrently with the converse rules of differentiation, is enabled to put him early in possession of many useful results of the science. After some problems in maxima and minima, the author explains the mechanical genesis and the leading properties of the conic sections; and shews the method of obtaining, by the integral calculus, the areas of surfaces and the volumes of solids. The student is then made acquainted with the properties of logarithms, (illustrated by the logarithmic curve,) with the differentiation of exponential and circular functions, and with Maclaurin's and Taylor's theorems. The properties of the cycloid are investigated, and an appendix is added, tracing the rise and progress of this department of the mathematics. The answers to the examples are given at the end of the book, in order that they may be detached if desirable.

V.

ELEMENTS OF ARITHMETIC.

By AUGUSTUS DE MORGAN, Professor of Mathematics in University College, London. Fifth Edition, with Eleven new Appendixes. Royal 12mo. 5s. cloth.

VI.

A FIRST BOOK ON PLANE TRIGONOMETRY.

Geometrical Trigonometry, and its Applications to Surveying, with numerous Examples, for the Use of Schools. By G. W. HEMMING, M.A., Fellow of St. John's College, Cambridge, and Author of a Treatise on the "Differential and Integral Calculus." 12mo. With Diagrams. 4s. 6d.

VII.

THE ELEMENTS OF STATICS, DYNAMICS, AND HYDROSTATICS,

With an Appendix on the Laws of Light, the Formation of Images by Lenses, and the Nature of Sound. By SAMUEL NEWTH, M.A., Fellow of University College, London. Royal 12mo. boards. 6s.

VIII.

AN ELEMENTARY TREATISE ON MECHANICS,

For the Use of the Junior University Students. By RICHARD POTTER, A.M., late Fellow of Queen's College, Cambridge, Professor of Natural Philosophy in University College, London. Second Edition. 8vo., with numerous Diagrams. 8s. 6d. cloth.

Made in the USA
Monee, IL
04 November 2026

67385015R00134